The Design of Interior Circulation

People and Buildings

The Design of Interior Circulation

Peter Tregenza

BArch, MBldgSc, PhD, RIBA
Department of Architecture
University of Nottingham

VNR **VAN NOSTRAND REINHOLD COMPANY**
NEW YORK   CINCINNATI   ATLANTA   DALLAS   SAN FRANCISCO

Van Nostrand Reinhold Company Regional Offices:
New York   Cincinnati   Atlanta   Dallas   San Francisco

Van Nostrand Reinhold Company International Offices:
London   Toronto   Melbourne

First Published in Great Britain 1976 by Crosby Lockwood Staples

Copyright © 1976 by Peter Tregenza

Library of Congress Catalog Card Number: 76-10846
ISBN: 0-442-80060-6

All rights reserved. No part of this work covered by the copyright hereon may be reproduced or used in any form or by any means – graphic, electronic, or mechanical, including photocopying, recording, taping, or information storage and retrieval systems – without permission of the publisher.

Published by Van Nostrand Reinhold Company
450 West 33rd Street, New York. N.Y. 10001

15 14 13 12 11 10 9 8 7 6 5 4 3 2 1

Library of Congress Cataloging in Publication Data

Tregenza, Peter.
   The design of interior circulation.

   Bibliography: p.
   Includes index.
   1. Pedestrian facilities design.   2. Buildings – Mechanical equipment.   I. Title.
NA2543.P4T73   1976   729'.25   76-10846
ISBN 0-442-80060-6

Printed in Great Britain

Preface

In this book I have collected some of the information that is needed in designing for the movement of people in buildings. It begins with a discussion on the effect of the overall form of a building on the nature of the interior circulation; then there are two chapters on lifts and other mechanical systems; the fourth chapter covers the design of stairways, corridors and pedestrian areas in general; and two appendices contain graphs and tables for passenger lift calculations. The book is written for those concerned with the design of large buildings, principally practising architects and those consulting engineers who, in one field or another, deal with pedestrian movement. Its scope is limited mainly to aspects of design which are readily quantifiable but I certainly do not intend to imply by this that I consider other topics unimportant. Far from it: my purpose is just to summarise for the designer the technical basis of the work.

Although some sections of the book are based on research projects in which I have been involved, most of it is derived from other sources. These vary widely. On the topic of pedestrian movement in stairways and corridors, for instance, there are on the one hand many reports of surveys which show that crowd behaviour is complex and our knowledge of it far from complete; on the other hand there are statutory codes on means of escape from fire which impose rigorous constraints on the design of a building. I have attempted to employ usefulness to the designer as the criterion for selecting material for inclusion but this was not always straightforward. Differing survey methods and variations in presentation have in some cases made comparison of research results difficult, and to furnish the designer with information he can use directly in initial calculations I have selected out of imprecise or inconsistent data the values that I consider most likely to be correct. No doubt revision will be necessary as more research is published. Where tables contain such arbitrary values I have indicated this with the word 'approximate' in the heading. The text contains the theory on which calculations are based, where this material has not been collected elsewhere. Metric units are used throughout; in many cases the values given have been converted from other units.

Without making the book long and very quickly out of date it would not be possible to describe all relevant statutory regulations and codes of practice. I have referred quite extensively to British publications, and in some instances linked these with requirements in other countries, but there is legislation covering safety in several aspects of interior circulation and the designer must be familiar with local requirements.

I have not discussed comparative costs of mechanical equipment. In general there is much that is commonsense (even though ignored sometimes) and a fair discussion of the subject can be found in some of the books listed in the bibliography. It would have been useful to have given actual prices of the main mechanical units but, alas, during the period in which the text was prepared the interval between successive price changes was much less than the time needed to publish a book.

P. R. Tregenza
January 1976

Errata

page 54, para. 2.4.2	for $\left(\dfrac{m}{k}\right)$ read $\binom{m}{k}$
page 94, line 12	for $-0{\cdot}5$ read $-0{\cdot}05$
page 133	for E_h read t_h

Contents

Preface v
Principal symbols ix
List of tables x
List of illustrations xi
Acknowledgements xii

1 Introduction
   1.1 Interior circulation and building form   1
   1.2 Circulation routes   6
   1.3 Detailed planning   11

2 Lifts and escalators: traffic-flow calculations
   2.1 Preliminary selection of lifts   15
   2.2 Lift-performance calculations   22
   2.3 Models of lift performance   35

       2.3.1 Upward traffic, constant number of passengers in lift cars on departure   36
       2.3.2 Upward traffic, constant probability of a call   37
       2.3.3 Random inter-floor traffic, constant probability of a call   39
       2.3.4 Variation in cycle time   40
       2.3.5 Bunching of lift cars   43
       2.3.6 Doors re-opening   44
       2.3.7 Digital simulation   45
       2.3.8 Choice of model   51

viii Contents

2.4	Capacity of paternoster lifts	53
	2.4.1 Capacity limits	53
	2.4.2 Random inter-floor traffic	54
2.5	Capacity of escalators	59

3 Mechanical equipment

3.1	Traction and electro-hydraulic lifts	63
	3.1.1 Overall dimensions	63
	3.1.2 Machine-room equipment	66
	3.1.3 The car assembly	72
	3.1.4 Suspension	77
	3.1.5 Operating systems	78
	3.1.6 Building construction work	81
	3.1.7 Special requirements	84
3.2	Paternoster lifts	85
3.3	Escalators and passenger conveyors	87

4 Stairs and corridors

4.1	Empirical calculations	91
	4.1.1 Corridors	92
	4.1.2 Doorways	100
	4.1.3 Stairways	100
	4.1.4 Waiting areas	103
4.2	Escape from fire	105
4.3	Special requirements of the disabled	113
	4.3.1 Corridors and waiting areas	114
	4.3.2 Stairs	115
	4.3.3 Ramps	116
	4.3.4 Handrails and barriers	116
	4.3.5 Doors	117

Bibliography		119
Appendix 1	Tables and graphs for selecting lifts	132
Appendix 2	Tables of E_h, E_{s1}, E_{s2}, E_{ss} and E_p for unidirectional traffic and random inter-floor traffic	139
Index		157

Principal symbols

kg	kilograms
m	metres
min	minutes
P	persons
s	seconds
a_c	number of persons carried in unit time by a lift system of c cars
α	arrival rate of persons
c	number of lift cars in a system
d	distance
E_h	expectation of the highest floor reached by a lift car during one cycle
E_p	expectation of the number of passengers carried in a lift car during one cycle
E_s	expectation of the number of stops made by a lift car during one cycle
E_{s1}, E_{s2}, E_{ss}	expectation of the number of stops, during one cycle, preceded by 1, 2 and 3 or more storeys of non-stop travel, respectively
E_w	expectation of passenger waiting time
f	acceleration
J	rate of change of acceleration
n	the highest floor served by a lift system, the floors being labelled 0, 1, 2, ... n upwards
P_r	the probability of r persons arriving to travel from one given floor to another during one cycle of the lift
R	one of a series of random numbers
t_h	time taken by a lift car to travel one storey distance at full speed

t_p mean total of the time taken by a passenger to enter and to leave a lift car

t_{s1}, t_{s2}, t_{ss} door operating time at landings plus the difference in time between journeys of only 1, 2 and 3 storeys respectively and the car travelling at full speed over the same distance

v velocity

List of tables

2.1 Mean interval between successive departures of lift cars.
2.2 Nominal traffic values for peak-loading calculations.
2.3 Lift behaviour relative to passenger arrival rates.
2.4 Simulation of lift-call patterns.
2.5 Number of passengers in a paternoster car on its departure from a landing.
2.6 Approximate capacity of escalators.
4.1 Approximate mean walking speeds. Corridor capacity.
4.2 Approximate reductions from the effective width of a walkway.
4.3 Minimum width of straight corridors.
4.4 Approximate capacity of openings.
4.5 Approximate mean speeds of movement up stairways. Stair capacity.
4.6 Occupant load factors.

List of illustrations

1.1 Variation of services and circulation area with building height in a sample of ten office buildings.
1.2 Rooms along a single-sided corridor.
2.1 Growth of queues with four 2·5 m/s lifts serving morning peak traffic in an eleven-storey building.
2.2 Probability of there being P passengers or fewer wishing to enter a lift car.
2.3 Variation of inter-departure interval with arrival rate.
2.4 Output of a typical lift-calculation program.
2.5 Down peak traffic: probability of a lift car being full.
2.6 Sensitivity of lift-capacity calculations.
2.7 Effect on cycle times of imposed delays to lift cars.
2.8 Simplified flow diagram of a lift-simulation program.
2.9 Variation of paternoster capacity with building height.
2.10 Variation of escalator capacity with speed.
3.1 (a) Overall view of an electric traction lift; (b) Overall view of a paternoster system.
3.2 A lift machine room with geared variable-voltage equipment.
3.3 An electro-hydraulic lift.
3.4 A standard lift car in the factory.
3.5 Paternoster drive equipment before installation.
3.6 An escalator during installation in a department store.
4.1 Observed walking speeds in an indoor shopping mall.
4.2 Variation of mean walking speed with crowd density.
4.3 Principal limits on travel distance in tall office buildings.

Acknowledgements

I am grateful to the publishers of *The Architects' Journal*, *Architectural Science Review* and *Building Science* for permission to reproduce material from articles of mine originally printed in these journals; to the London Transport Executive and the Council of the Institution of Civil Engineers for the material illustrated in Figure 2.10; to Hammond & Champness Ltd for Figures 3.2 and 3.3; to Otis Elevator Company for Figures 3.4, 3.5 and 3.6. I am also indebted to the publishers of *Building* for sponsoring an RIBA Research Award for work on the prediction of lift performance; to many of my colleagues at Nottingham University for constructive criticism and other help; to Mrs Anne Widdowson for her efficiency in preparing the typescript; to my wife for her continuing help.

P.R.T.

CHAPTER
1
Introduction

1.1 Interior circulation and building form

It is primarily the basic shape of a building that determines the efficiency of its internal circulation system. Buildings of the same internal floor area which differ in height or in dispersion over a site show significant variation both in the proportion of space taken by elements of circulation and in the monetary costs of lifts and other mechanical equipment. The time required by a person to travel from one part of the building to another and the interest or enjoyment of doing this are affected, too, by the basic form that is chosen. Although the detailed design of stairs, corridors and mechanical systems is important, and is the subject of the greater part of this book, the quality of the building depends at first on the basic analysis of the flow of people that will occur and the consequent decisions on the fundamental form of the structure.

One example of the relationship between the height of a building and the proportion of the interior space that is taken by circulation and other services is illustrated, with a sample of ten office buildings, in Fig. 1.1.[145] The vertical axis of the graph gives the percentage of the internal floor area that is occupied

2 The Design of Interior Circulation

by elements of horizontal and vertical circulation and by service spaces; the horizontal axis shows the corresponding number of storeys. It can be seen that the proportion of floor area that is not directly profitable tends to increase with the height of the buildings and, with this sample, approximately doubles between three storeys and fifteen. This relationship between

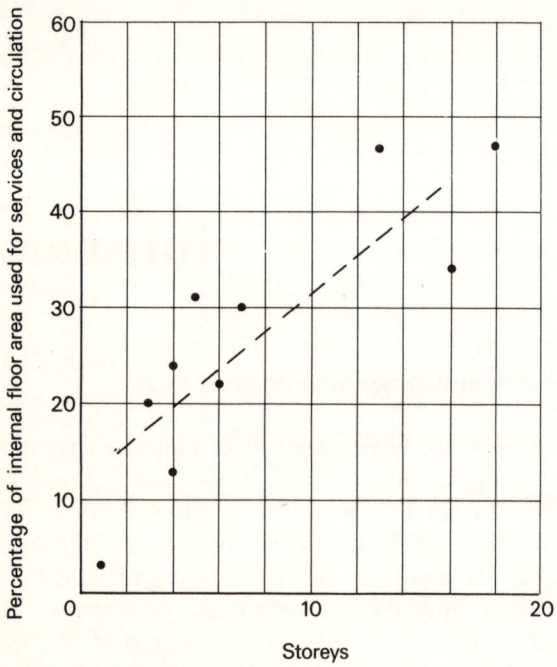

Fig. 1.1 Variation of services and circulation area with building height in a sample of ten office buildings

profitable area and building height is found to be non-linear in samples that include buildings much taller than twenty storeys. The graph becomes flatter with the higher buildings, a result probably due partly to lift zoning practice and partly to the tendency for taller buildings to be generally larger.

Alternative designs for a single building can exhibit an even greater variation of circulation area with height than shown in Fig. 1.1. It occurs because the total plan area effectively occupied by any vertical shaft in the building increases in proportion to the number of storeys through which it passes. There are two principal elements concerned: emergency escape stairs and lifts. Dimensions of the former are specified in codes of practice governing means of escape from fire. Increasing the number of storeys, keeping the intended population of a building constant, may under some codes allow a reduction in the plan area of each stairway; therefore a doubling of the number of storeys tends to cause slightly less than a two-fold increase in the total area of stairs. An increase in building height leads, on the other hand, to a greater number of lifts and if the design for an eight-storey building were replaced by one of sixteen storeys the proportion of area occupied by lift wells would be more than doubled. The total area used for ducts, for the structural framework, and for lavatories and other service rooms can also become significantly greater in taller buildings.

With each increase in the proportion of total plan area used for circulation and services there is a corresponding increase in the real cost of each square metre of directly profitable space (quite apart from the tendency of construction costs to be greater for a tall structure than one that is low).[37, 145] So if there is a cost limit to a new project there is a resulting limit, in design, on the slenderness of the building. Beyond this an adequate circulation system is impossible, long waiting times for lifts usually reflecting any deficiency. In a low structure extended across the site there is, too, a tendency for the area devoted to circulation to be increased. The effect may be less pronounced, for space used in horizontal circulation can be used for other purposes and, except for long routes within transport terminal buildings, mechanical systems are uncommon. But in general the cost of internal circulation, whether measured as the values of mechanical equipment and internal floor area, or related to the time spent by people moving within the building, is lowest when the form of a large building is a low compact block. Then the external surface area is at a minimum, and so are the radial distribution distances of fluid and electrical services. It is normal to find that the total lifetime costs of a large building are lowest when it takes a regular and compact shape. Variance from this form follows the introduction of other values: high unit site prices leading on the one hand to tall buildings; visual criteria, and problems of access and noise, leading on the other hand to dispersed forms.

Fig. 1.2 Rooms along a single-sided corridor

Determining the relative locations of individual rooms is the other aspect of basic planning. Figure 1.2 shows a sequence of rooms along a single-sided corridor, giving a one-dimensional form of the problem and illustrating its nature. The width of each room is known, and so are the rates of traffic flow between every pair of rooms. The object is an arrangement of rooms in which the summation of the product of the distance between each pair of rooms and the rate of movement between them is as small as possible. That is, a minimising of

$$C = \sum_{i=2}^{m} \sum_{j=1}^{i-1} t_{ij} d_{ij} \qquad (1.1)$$

where the rooms are labelled 1 to m; t_{ij} is the traffic flow between two rooms and d_{ij} is the travel distance. C is a measure of the circulation cost. Simmons[134] has described a technique for solving the one-dimensional form of the problem.

It is not only the amount of traffic generated by any room that affects its optimum location. Unless the large room labelled C in the illustration is the source of a relatively intense flow of traffic, the cost of circulation within this group would be reduced by placing the room at one end of the line. The larger a space, the longer the journeys between the rooms around it. In the best layout of a group of rooms there tend to be in the centre small rooms generating heavy traffic; around these, larger rooms with strong circulation links and small rooms with lesser traffic; on the perimeter, large rooms having only weak links with the others. Costs of circulation are increased when rooms with heavy mutual traffic are separated, or areas requiring public access are not at the ground floor of a building. A roof-top restaurant or a second-floor cinema may represent the best use of the resources available, but the increased cost is significant and must be balanced by the increased value of the accommodation.

Main entrance areas, being small but carrying a major quantity of traffic, often become principal nodes of a circulation network, with the optimum shape of the building dependent on the ratio between internal traffic and movement to and from the outside.[105] In the extreme case of commercial multi-storey structures, where the central core of lifts must carry the entire population of the building to the ground floor within a short period, the extent to which the fixed elements of the ground floor constrain the location of the lifts core and the escape stairs may determine the feasible plan forms, and hence the profitability, of upper floors.

The mathematical problem becomes more complex when extended from a single line of rooms to two or three dimensions. The shapes of rooms and their fitting together are one factor of this, the other is the need to define circulation routes. In a large multi-cell building the pattern of circulation routes tends to be a simple rectilinear skeleton, as this reduces the total space used for circulation although at the cost of increased travel distances, so the problem of designing involves both the selection of the shape of the circulation spine and the location of rooms around this.

The obvious strategy for solution, the comparison of all possible arrangements, is limited by the number of combinations that is possible. There are 720 alternative sequences of the six rooms in Fig. 1.2, and with m rooms there would be $m!$ possible arrangements. This is halved if lateral symmetry of the row is taken into account, and other criteria for the location of rooms in a real building also reduce the number, but when m represents the number of rooms in a building of even moderate size the number of possible different plans is very large.

Several methods have been proposed for allocating the relative locations of rooms, in two and three dimensions, for a minimum circulation cost. (See, for instance, Refs. 9, 86, 90, 105, 132, 134, 161.) Most of the techniques are written as computer programs, and most of those which are entirely automatic methods make major simplifications to the problem – the division of large rooms into small elements that can be formed into different shapes, and the placing of rooms in space without consideration of the actual circulation routes. But methods which give just an allocation of the elements within a building for a particular measure of circulation cost are valuable only in special cases. Normally there are two conditions that reduce their applicability:

(1) The data are not precise. The rates of traffic flow between every pair of rooms throughout the lifetime of the building are simply not known.

(2) The form of building that gives a minimum cost of circulation is not necessarily the best when other criteria are considered. The prerequisites of an efficient structural framework or of economical distribution of services may be at least as stringent. Interactive computer techniques, those in which the operator can modify partial solutions to introduce previously unspecified factors, are, for this reason, likely to be the most practical of the methods developed.

1.2 Circulation routes

The various elements of circulation within buildings – lifts and escalators, stairs and corridors – vary in passenger-carrying capacity and their spatial and structural requirements. The designer is required to match these characteristics with those of the traffic generated.

Conventional lifts. A conventional lift system has a number of cars, in separate vertical shafts and mechanically independent, that can carry groups of people at relatively high speeds. The maximum car speed that can be attained is directly related to the distance to be travelled between stops, and journey times are at a minimum when cars can travel between two floors with no intermediate calls.

Each additional stop causes a significant increase in journey time. The car loses speed as it approaches the landing, there is a delay while the doors operate, and several metres of travel are required for the car to reach maximum speed again after departure; and it may not reach this if required to stop shortly at another landing. The total increase of time may be 10 s, so three additional stops during a journey can add half a minute both to the time spent by a passenger in the lift and (with a single lift) to the time that potential passengers may wait. An increased period between successive departures of a lift from a landing leads to an increase in the average number of passengers in the car and thus to the probability of more stops in following trips.

The maximum rate at which people are carried over a given distance occurs when a group of passengers enters the car at one landing and is taken to a single destination. There is a decrease in efficiency with traffic such as that of the morning peak in commercial buildings, the passengers all entering at the ground floor but travelling to various higher storeys. With random inter-floor traffic, passengers arriving unpredictably to make journeys between any landings, the number of people carried by a conventional lift in unit time is small. In a building of ten floors above ground level a group of three 1·5 m/s lifts might have the capacity to carry 25 P/min directly from the ground to the highest floor, 11 P/min with people entering the cars at the ground floor and leaving randomly at higher landings but only 5 P/min with completely random inter-floor traffic, in each case the average interval between successive departures of a lift being the same.

To maintain efficiency, constraints must be imposed on the overall planning of lift systems:

(1) Lifts serving a single route should be grouped together. This reduces waiting times at landings. Continuing the example above, three separate lifts independently taking 3·7 P/min might each have an interval of 80 s between departures from ground level. The three lifts grouped, operating together at 11 P/min, would yield a mean interval of 30 s, a reduction in the average waiting time of about 60 per cent. Entrances to the lifts must be close to each other so that a group of passengers is able to move into the first car that arrives, without increasing the

period that a car remains at the landing. Two rows of lift doors, facing each other, should be 2·5–3·5 m apart, and there should not be more than four lift entrances alongside one another. The British Standard Code of Practice, CP 407:1972,[23] illustrates typical lift lobby arrangements. Marmot and Gero[97] have derived empirical relationships between lobby areas, the number and grouping of lifts, and the internal floor areas of office buildings. In the survey of twenty buildings that formed the basis of the study it was found that the lobby areas varied widely; the mean area of a ground floor lobby was 18·2 m^2 per lift in systems with one upper zone and 35·4 m^2 per lift when the upper storeys were divided into two or more zones. The mean lobby area on each upper floor was approximately 5·6 m^2 per lift.

(2) The control systems of the lifts must be linked. This implies that at intermediate landings there are effectively only two call buttons, 'up' and 'down', whatever the number of cars, and there is only a single button at each terminal floor. When a button has been pressed the call is answered by the next available car and then cancelled automatically, avoiding unnecessary stops by other cars.

(3) At the base of a tall building it is common to have entrances at several levels, from car-parking areas and different levels of street access. If passengers are brought to one main loading floor to take the lifts, the number of stops made by lifts during peak traffic flows is reduced, and hence the number of lifts required is smaller. The use of secondary lifts or escalators, to bring passengers to the main terminal floor, can be economically justifiable.

(4) The total volume of lift shafts and equipment in a very high building is reduced when different groups of lifts serve separate zones of upper floors. In a building of thirty storeys, one group of four lifts might join the ground floor with the upper fifteen floors, and another group of four lifts be installed to serve only the lower fifteen. Zoning reduces the convenience of movement between floors above ground level; it is unusual to find it in buildings of less than fifteen storeys occupied by a single organisation, but is almost universal in commercial buildings of thirty storeys and more.

Paternoster lifts. The paternoster is a chain of small lift cars moving continuously. At each landing two doorways show a series of cars moving past in opposite directions; successive cars are separated in distance by the height of one storey. The cables carrying the cars pass over large pulleys at the top of the system where the cars reverse and travel down the adjacent shaft. Another pair of pulleys guides the cars at the bottom of the shafts. A paternoster system is illustrated in Fig. 3.1(b).

Passengers must step into moving cars. The speed is low in comparison with conventional lifts, usually between 0·3 and 0·4 m/s, but even this limits the use of paternosters to able-bodied people and generally precludes their installation in buildings with general public access.

The cars are small, with a maximum capacity of two people, and there is time for only two passenger movements at each landing — for two people to enter or leave the car, or for one person to move in each direction. The resulting traffic characteristics differ from those of conventional lifts:

(1) Long journeys are slow. Moving through ten storeys may take 2 min, and a steady flow of passengers travelling the full height of the system reduces the number who can enter at intermediate floors. With cars spaced 3 m apart travelling at 0·3 m/s a maximum of 12 P/min can be carried upwards from the lowest floor and a similar number from the highest floor, but if the passengers travel the full height of the building none can enter at any other floor.

(2) The capacity is higher with random inter-floor traffic. In a building with ten upper floors, and the mean traffic rate the same between all floors, a total of 35 P/min can be carried, with very short waiting times at landings. This is about ten times the capacity of a conventional lift system of similar plan area.

Escalators and passenger conveyors. Under conditions where a queue forms at the foot of an escalator, 150 P/min can be carried on equipment with a stair tread 1 m wide. The first situation in which escalators are employed is, therefore, one in which very heavy traffic must be carried a relatively short vertical distance. Except in transit buildings it is unusual to find a single moving stair rising more than 6 m, since the large volume required and the horizontal separation of entry and exit points can make this unsuitable. But low speed and absence of an enclosing shaft give a second type of use. This is in shopping centres and other buildings where display is important. In a department store escalators can be used to attract shoppers past merchandise on a sales floor, and goods can be exhibited to upward and downward moving passengers. A casual shopper tends to be reluctant to use stairs or lifts, but much less hesitation is found with escalators. The profitability of the upper floors of a large store can be dependent on the use of this equipment.

Passenger conveyors, or moving ramps, share many of the characteristics of escalators. The major difference is the angle of slope that is possible. Standard angles for escalators are 30° and 35°; a moving ramp can be installed at any angle up to 12°, at which it becomes uncomfortable for pedestrians. They are essentially devices for carrying people over substantial distances and are uncommon in buildings other than airports and large shopping centres. There is a secondary use in moving heavy flows of people through major displays, as in World Fairs or popular permanent exhibitions. The passenger is carried slowly and with excellent vision through the display, but he is unable to stop, and the exhibits must be designed for this type of viewing.

Both escalators and passenger conveyors can be reversed, to carry peak flows in opposite directions, and they have the additional advantage that on mechanical failure they continue to be usable routes, though of reduced capacity.

Stairs and corridors. Stairways, in relation to their plan area, can carry heavy flow rates; with only moderate crowding a one-storey stair 2 m wide can be entered by 60 P/min. Stairs are tiring but a simple limit of height cannot be defined, for tolerance of long travel distances depends, among other factors, on the age-group of the building's occupants, their motivation and the frequency of use. Distances considered normal by a crowd entering a stadium or a theatre gallery, for instance, exceed current practice in housing and commercial buildings where a four-storey flight of stairs is often taken to be the maximum for normal use.

Staircases are used as principal escape routes in multi-storey buildings and it is necessary, when designing, to distinguish between stairs that are intended for this purpose and those only for normal use. In the latter case the dimensions may not be determined primarily by the traffic flow, and the view from the stairs and the sense of progression through the building can be major factors of the quality of the design. Escape stairs, on the other hand, must take specified dimensions and, with few exceptions, must be enclosed in a ventilated shaft with walls of known fire resistance.

The distinction between normal and emergency use can be made also with horizontal routes. The paths of random circulation may not initially be defined as such and will alter – with changes in the use of the building, with the location of furniture and temporary partitions, and with many minor conditions. The volume of pedestrian traffic might not be a major criterion. But where escape from danger is one of the functions the character is changed. The route must be obvious, not just to regular users but to the stranger, the young, the infirm and handicapped. The distance to safety from heat and asphyxiation must be related to the rate at which a fire could increase, and the dimensions of the corridor proportionate to the number of the users.

Provision for fire escape is mandatory, but the requirements vary with the governing authority and with the use, form and construction of the building. This is discussed in Chapter 4.

1.3 Detailed planning

The provision of circulation routes with characteristics that match the pattern of traffic is not, in itself, enough. Actions of individual people can affect the movement of others without it being obvious to the general user of a building, and procedures that appear to give immediate benefit can increase inconvenience elsewhere. An example of this is the habit of pressing a lift landing button and then, if a car does not arrive immediately, using the stairs or paternoster for the journey; such practices that cause a lift car to make unnecessary stops increase the overall mean waiting time. The detailed planning of the building can affect the way in which the different elements are used, and it is possible to reduce inefficient operation of mechanical systems and unnecessary impedance to movement in corridors and stairways.

Separation of different traffic flows. Unless, in a tall building, interchange of passengers from one system to another is intended for regular journeys, there is rarely a need for the different elements of vertical circulation to be alongside one another. Having a single 'circulation core' does not necessarily yield efficient patterns of movement, and troublesome interaction of the type described is reduced when lifts are separated from the other systems. If it is intended, for example, that passengers travelling several storeys should take conventional lifts and those making shorter journeys should use stairs or paternoster this should be made clear. It can be achieved by placing the lift shaft further from the origin of traffic than the other route, making the more efficient system the more convenient. Similarly, if separate groups of lifts travel to two or more different upper zones, confusion and mistakes at the ground floor can be avoided by ensuring that the different queues are separate and making obvious the identity of each group. In corridors and stairs, the area required for each person is greater when the traffic is mixed in direction and speed than with a single flow. Although it is rare to need separate routes for opposing flows, when heavy traffic occurs in a limited space some channelling is warranted.

Movement of trolleys, wheelchairs and bulky goods hinders heavy pedestrian flows, and is itself hindered, and where there is regular traffic of this nature specific provision for it is necessary.

Separation of waiting areas. Even two or three people waiting for a lift can cause a significant reduction in the rate of flow down a busy corridor, and those who are waiting need to see lift indicators and be free to move quickly to any of several lift openings. When there is a heavy flow of traffic, space must be allocated for people who are not part of this flow, whether queuing for lifts, reading direction boards or for other reasons.

Provision of information. Although strangers in a public building, uncertain of their route, can hamper other traffic, it could be held that to give inadequate direction is discourteous to the individuals involved. Visitors to almost any building can include those with physical and mental handicaps and their requirements should be considered during the design process.

A stranger, entering a building, is usually directed by one or more of three procedures: observing and following other people in the building; reading signs or notices; asking another person. Visitors frequently exhibit shyness and, to the diffident, the three methods are in order of preference. If a queue can be joined or an obvious route followed with little delay, this is an advantage. But behaviour varies and, although there is much clarity that the architect can achieve in the planning of the circulation and in the design of signs, where personal interaction is a factor the attitudes and behaviour of the staff in the building may be dominant.

Lifts are complex pieces of mechanical equipment and in some types of building even the regular users of them may need some instruction. Not all malpractices can be prevented by the designer. The main points to be understood are the effect of delaying lift cars, and the action of the safety equipment. It should be clear that to delay a lift or to cause one to stop unnecessarily has the effect that other people have longer to wait. Actions that are typical include pressing both 'up' and 'down' call buttons at landings (which normally gives no advantage), the holding open of doors with trolleys or parcels during a short visit to an upper floor, and the unnecessary use of lifts for short journeys during peak arrival and departure periods. Unfortunately, behaviour that gives a short-term advantage is more common when the capacity of the lift system is inadequate. This aggravates the problem, but when the anticipated waiting time is very long the temptation to shorten it becomes stronger.

The safety devices of a lift are designed to protect passengers in the car when there is a mechanical failure or an interruption of the power supply. In these circumstances, which should be infrequent, the car can stop in darkness between floors; although this can be distressing, accidents that cause personal injury are very rare. But fatal accidents to people climbing within the lift shaft are recorded in number. They have occurred during maintenance of the machinery and to passengers forcing an exit when a car starts unexpectedly after a halt. Every lift car is equipped with an emergency call device, a telephone or an alarm bell, and lift users should be familiar with its use, just as there must be an adequate procedure for responding to it. Passengers should be instructed to remain in a stranded car.

Finally, there is information that may be required by the architect's client or the managers of a building. The design of an internal circulation system is based on a number of assumptions about the use of the building and the resulting patterns of traffic. It is analogous to the adoption of specific loading conditions in the design of the structural framework. Changes in the use of the building can cause a failure of the system, either a mechanical breakdown or crowding and long waiting times for the occupants. This may occur without any increase in the number of people using the system. In an educational building the academic timetable can be planned in a way that generates very high peaks of traffic flow, and alteration of times that different groups of workers in an office building begin or end their working day may affect significantly the maximum demands on the lifts. It is unlikely that the best allocation of resources for the building as a whole lies with the design of the circulation to give short waiting and journey times under extreme conditions, and it is in some cases necessary to discuss with the client the initial constraints to be adopted, as well as to provide, when construction is completed, details of the loading assumptions and physical limits of the mechanical equipment.

CHAPTER
2
Lifts and escalators: traffic-flow calculations

2.1 Preliminary selection of lifts

The principal measures of lift performance are, first, the number of passengers that can be carried in a given time and, second, the period that passengers must spend waiting for a lift and travelling between floors. Criteria based on these are not necessarily satisfied simultaneously. A system, for instance, in which a large lift car departs from the ground floor of a building once every 4 min might have a capacity substantially greater than the average number of passengers arriving, but the corresponding mean waiting time of at least 2 min would rarely be considered satisfactory.

Little research has been published on the factors that determine passenger satisfaction with lift systems. In effect, the standards generally adopted are based on market demands, the results of comparison between the value of service given to lift users and the considerable capital cost of large lift systems. Table 2.1 lists criteria used in current practice. There is no reason to believe that lifts that conform to these prove unsatisfactory to the users, but neither can it be assumed that changing assessments of the economic balance between passenger expectations and equipment costs will not cause these standards to vary in time.

The values given are those to be actually attained in the installation, inter-departure times that should be exceeded only in a very small proportion of journeys. If one or more of a group of lifts is likely to be out of service frequently – for maintenance work, through vandalism or for carrying goods – this must be taken into account.

Table 2.1. **Mean interval between successive departures of lift cars**[2, 23, 95, 100, 137]

	Interval (s)
Offices and other commercial buildings, industrial buildings	25–35
Leisure buildings, multi-storey car parks	40–50
Department stores	30–50
Hotels	30–60
Housing	60–90

An interval at the lower end of each range tends to be associated with buildings having a high standard of finishes and fittings.

Knowledge of the average period between successive departures of a lift car is not, in itself, sufficient to predict the actual waiting times of passengers. If people arrive randomly at the lift lobby, the mean waiting time is half the interval between departures only if the interval is constant. Any variation causes an increase in the mean waiting time, and it is possible for this to be greater than the mean interval between departures. Associated, then, with the values in Table 2.1 is the requirement of regularity in the departures of cars under heavy traffic conditions, although this may imply redundant capacity in the system.

The demand on the lift system in a building varies throughout the day. It is necessary to forecast the duration and intensity of peak traffic flows and to base the design on these rather than on the average flow over a longer time. Figure 2.1 illustrates the results of simulating lift performance under an arrival rate of passengers that becomes substantially above lift capacity. The number of passengers arriving per minute

Lifts and Escalators: Traffic-flow Calculations 17

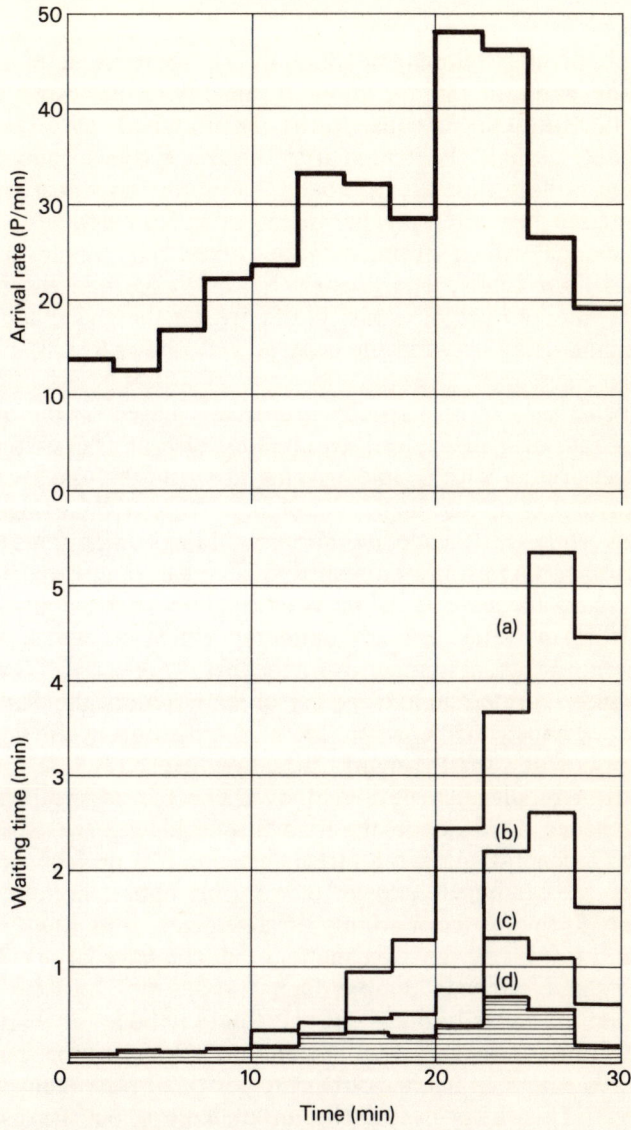

Fig. 2.1 Growth of queues with four 2·5 m/s lifts serving morning peak traffic in an eleven-storey building
(a) 900 kg (12 P) cars (d) 2100 kg (28 P) cars
(b) 1200 kg (16 P) cars Results from computer simulation
(c) 1500 kg (20 P) cars

increases to a pronounced peak and then declines sharply, typical of traffic that may be found during the morning in a city building, and although the nominal capacity of the system with four 1200 kg cars is equal to the mean arrival rate over the half-hour period, the irregularity of the arrivals causes an average waiting time greater than $2\frac{1}{2}$ min for those passengers who arrive just after the period of maximum arrivals. This corresponds with a queue of about sixty-four people in the ground floor lobby. The greater the difference between peak arrival rate and actual capacity, the greater the queue and the longer the delay between the peak arrival rate and peak queue length.

Calculations of lift capacity are usually based on the mean arrival rate over the central five minutes of peak traffic flow. It has been found, with people arriving at work, that the average proportion of a building's population to arrive during the busiest 5 min is the same in different buildings of the same type when other factors are comparable. Typical values that have been found adequate as the basis of morning peak calculations are listed in Table 2.2, although the extent to which they represent the actual number using the lifts is not certain. On the one hand, absenteeism and the use of stairs reduce the demand for lift service; on the other, there is reduction in the actual capacity of lifts for the main traffic flow caused by the use of lifts for irregular journeys and by the needs of mechanical maintenance. On balance, the figures in Table 2.2 are satisfactory as a general guide, but factors specific to a project should always be examined: staggered working hours, policies on arrival discipline, car-parking arrangements, and links with public-transport systems are matters that can have large effects on the rate of arrivals in a building.

When the capacity of a system is found to be lower than the initial demand, workers tend not only to use stairs more but to adjust their arrival habits so that the period of peak demand is altered.[39] This is not necessarily unsatisfactory, but the degree to which the arrival pattern in a building with a poor lift service differs from that of a similar building in which the lifts have adequate capacity may be taken to be a measure of dissatisfaction.

Table 2.2 Nominal traffic values for peak-loading calculations [2, 23, 95, 100, 138]

	Number of people arriving in 5 minutes, given as percentage of building occupants accommodated above ground floor level
Single occupancy office buildings, industrial buildings	15–20%
Multiple tenancy offices	11–15%
Leisure buildings, hotels	10–15%
Housing	5–7%

There are several building types for which an overall estimate of this kind is inappropriate. Schools and universities, auditorium buildings and others where the internal traffic is formed in part by people entering and discharging from places of assembly require an analysis based on the anticipated use of the accommodation. The passenger flow in hospitals, airports and buildings such as courthouses is composed of several independent streams, some of which must be isolated, and a separate analysis of each is necessary. In multi-storey car parks the demand for lifts can be deduced from the maximum rate at which cars can enter and leave, and in other transport buildings an analysis of the road or rail traffic into the terminus is generally essential to an estimate of the rate of pedestrian flow.

The graphs of Appendix 1 are for selecting lifts during the initial design stages of a building. Graphs A and B describe lift performance when the passenger flow is upwards from the ground floor of a building, the morning arrival peak of commercial buildings. The number of upper storeys served by the lift is given on the horizontal axis of each graph, the mean arrival rate of passengers on the vertical axis. The first graph shows the arrival rate which, with each lift system, corresponds with a mean interval of 30 s between departures of lift cars; graph B gives the arrival rates corresponding with a 45 s interval.

It was assumed, in deriving the graphs, that passenger arrivals are random and independent of one another and that the departures of lifts from the ground floor are controlled to give only small variation in the inter-departure interval. The storey height was taken to be 3·3 m and it was assumed that the population of the building is evenly distributed between upper floors. Variation from this last assumption, an uneven distribution of passenger destinations, causes a reduction in the mean number of stops made by a lift, thus a shorter cycle time and greater flow capacity, but some patterns of loading can increase the variance of the cycle time and this tends to compensate for the lower mean. It should be noted that variation from the assumption of unidirectional traffic may cause a significant reduction in the capacity of a lift system. Under the intense flows of traffic at the beginning and end of a working day spasmodic journeys between upper floors, by cleaners or mail delivery for instance, can cause an increase in the round-trip time of lifts that appears disproportionate to the increase in traffic.

Example 2.1

An office building is to have ten floors above ground level, with a nominal occupancy of 800 people. It is to have a single tenancy but times of starting and finishing work will be staggered.

(a) Adopt 15 per cent of the nominal occupancy as the required 5-min capacity (Table 2.2), a mean arrival rate of

$$(800 \times 15)/(5 \times 100) = 24 \cdot 0 \text{ P/min}$$

(b) From the vertical axis of graph A it is found that, with a 30 s interval, at 24·0 P/min a car capacity of 1200 kg (16 P) is required.

(c) The point on the graph given by the intersection of the lines representing ten floors and 24.0 P/min lies between the curves of a system with four 2·5 m/s lifts and a system with four lifts of 1·5 m/s. The former would, under this traffic flow, have a mean interval of rather less than 30 s, the slower lifts an interval above 30 s.

If a mean interval of 45 s were permissible (which would be unlikely in a city office building) it can be seen from graph B that 1800 kg cars would be required and that three lifts of 2·5 m/s would be just adequate. Graph A shows that these lifts would give a 30 s interval at about 16 P/min (at which there would be an average of eight people in each car).

Tables for the preliminary selection of lifts for morning peak traffic are given by Williams [162, 163] and in CP 407.[23] Linzey,[93] Forwood[55] and other research workers have devised methods for computing the best combination of car size and speed when the traffic flow can be determined accurately. In most instances these are computer programs to calculate all feasible solutions for a given building and allow the cheapest to be selected.

Outgoing traffic from a building, particularly the peak flow as the working day ends, can be of greater intensity than any other pattern of traffic. All the occupants of a commercial building may leave within half an hour, and the total movement during the busiest 5 min can be substantially greater than the figures for up-peak traffic listed in Table 2.2. Greater waiting times appear to be tolerated and, in low- and medium-height buildings, far more people are willing to use the stairs. The round trip time of lifts is generally less than in peak morning traffic, provided cars have load-sensing devices that prevent full cars stopping unnecessarily, but in a large building it is essential that lifts operate under the type of control program in which ground floor delay is minimised and cars are allocated to intermediate landings. Given these factors, and a tendency for cars to be loaded to at least their nominal maximum capacity in the busiest period, a lift system that is satisfactory under morning peak conditions is usually also adequate under heavy outgoing traffic.

Graphs C and D allow the capacity of a lift system to be checked for random inter-floor traffic. These graphs illustrate lift performance where passengers arrive independently and randomly to take the lifts, and where the mean rate of travel is the same from each floor to every other. It is an extreme case: any change towards concentrating the traffic flow between fewer pairs of floors leads to a reduction in the number of stops made by a lift and thus, in almost all examples, to a shorter round trip time. When the total passenger flow is unidirectional traffic, to or from a terminal floor, the number of people in each lift car is at a maximum, the mean number being numerically one half the mean arrival rate (P/min) where the interval is 30 s. Random inter-floor traffic of the same total intensity yields longer cycle times but minimum car occupancy, for all the passengers are not in the lift car at the same time. These minimum values are shown in graphs C and D; when the anticipated traffic pattern is far from uniform, larger car sizes should be adopted.

Example 2.2

Continuing from the previous example: four cars of 1200 kg at 2·5 m/s have been selected. What is the capacity of the system under random inter-floor traffic?

From graph C, a mean interval of 30 s would occur when four cars at 2·5 m/s carry traffic with a total arrival rate of 11 P/min distributed over eleven storeys (ground level + ten upper floors). From graph D, a mean interval of 45 s would occur with an arrival rate of approximately 16 P/min.

Under random inter-floor traffic a lift may not stop at the same landing in successive cycles. The 'interval' is taken to be the period between successive possible departures of a lift from any floor, the period between departures if one or more passengers happen to be waiting at the landing on each occasion a car approaches.

2.2 Lift-performance calculations

We make the assumption that under morning peak traffic a lift car fills with passengers at the ground floor, travels to a number of upper floors where one or more passengers leave, reverses at the highest of these floors and returns without any intermediate stops to the ground floor. The time taken to complete a cycle from the initial departure from the ground floor to the passengers filling the car for the next trip is, then, a function of the following:

(1) The distance travelled by the car.

Travelling at maximum speed, the average time to cover a distance equal to the round trip would be $2E_h t_h$, where E_h is the expectation of the highest floor reached by the car (numbering the floors 0, 1, 2, ..., n upwards) and t_h is the time required to travel one storey distance at full speed, that is, the storey height divided by the velocity.

(2) The number of stops.

At each landing, additional time is required for the lift to brake and to accelerate again, and for the doors to operate. Over short journeys between adjacent floors, lifts with a high contract speed may not reach full velocity. It is convenient to express the component of the cycle time related to the number of stops as $E_{s1}t_{s1} + E_{s2}t_{s2} + E_{ss}t_{ss}$, where E_{s1}, E_{s2} and E_{ss} are expectations of the number of stops preceded by one, two and three or more storeys respectively, and t_{s1}, t_{s2} and t_{ss} the door operating time at landings plus the difference in time between journeys of only one, two or three storeys and the lift travelling at full speed over an equivalent distance. This gives a small, usually trivial, error in the calculation of very fast lifts that require more than three storeys to attain maximum speed and brake again; a method of calculating journey times is given in Chapter 3.

(3) The number of passengers carried.

If t_p is the average time required by a passenger to board the lift car and to leave it again at another landing, and E_p is the expectation of the number of passengers carried, the component of cycle time due to passenger movement is $E_p t_p$.

The mean cycle time (or 'round trip time') of a single lift is thus

$$T_1 = 2E_h t_h + E_{s1}t_{s1} + E_{s2}t_{s2} + E_{ss}t_{ss} + E_p t_p \quad (2.1)$$

The quantities E_h to E_p are listed in Appendix 2. They are tabulated against values of p_0, the probability of there being no passengers wishing to travel from one given floor to another during a particular cycle; increasing values of p_0 represent a decreasing intensity of traffic. With unidirectional traffic from the ground floor the probability of a lift car becoming idle on completing a cycle is p_0^n and the highest tabulated values of p_0

represent a probability of 0·2 or less that this will occur.
The corresponding passenger arrival rate, the number of passengers carried in unit time is

$$a_1 = E_p/T_1 \qquad (2.2)$$

Table 1 of Appendix 1 lists typical values of the lift-performance parameters, t_h to t_p. Taking these, or the values that describe a particular lift system, the relationship between the mean cycle time of the lift and the mean arrival rate can be calculated.

Example 2.3

A lift of 1·5 m/s contract speed serves the upper six floors of a building with unidirectional traffic from the ground floor. The storey height is 3·3 m and the lift parameters are those given in Appendix 1 for a 1·5 m/s geared v–v machine.

Using the table in Appendix B (unidirectional traffic, six floors above ground) the lift cycle times that correspond with various passenger arrival rates are obtained as in Table 2.3.

Table 2.3 Lift behaviour relative to passenger arrival rates

p_0	$t_h = 2\cdot2$ E_h $2E_h t_h$	$t_{s1} = 6\cdot8$ E_{s1} $E_{s1}t_{s1}$	$t_{s2} = 6\cdot8$ E_{s2} $E_{s2}t_{s2}$	$t_{ss} = 6\cdot8$ E_{ss} $E_{ss}t_{ss}$	$t_p = 2\cdot1$ E_p $E_p t_p$	T_1	a_1
0·50	5·0 22·0	1·8 12·2	0·8 5·4	1·4 9·5	4·2 8·8	57·9	4·3
0·55	4·8 21·1	1·5 10·2	0·7 4·8	1·5 10·2	3·6 7·6	53·9	4·0
0·60	4·6 20·2	1·2 8·2	0·7 4·8	1·4 9·5	3·1 6·5	49·2	3·8
0·65	4·3 18·9	1·0 6·8	0·6 4·1	1·4 9·5	2·6 5·5	44·8	3·5
0·70	3·9 17·2	0·8 5·4	0·5 3·4	1·3 8·8	2·1 4·4	39·2	3·2
0·75	3·5 15·4	0·6 4·1	0·5 3·4	1·2 8·2	1·7 3·6	34·7	2·9

Initially, one or two trials may be necessary to find the range of values of p_0 that gives figures of T_1 or a_1 in the required range. The size of E_p, the mean number of passengers carried per trip, is a useful guide in this. Table 2.3 lists those values that yield a cycle time of less than one minute.

When two or more lifts operate as a group the individual cycle times tend to be slightly greater than those of single cars carrying proportionate traffic. The practice of causing an automatic delay to the departure of lightly loaded cars, to prevent bunching, is one factor of this; the effect can be calculated when the variance of the natural cycle time is known (see Section 2.3.5). Normally, though, the accuracy of information on passenger arrival rates and lift-operating parameters does not warrant such precision and an increase of 5–10 per cent in the calculated cycle time yields a result with an accuracy well within that of the basic traffic assumptions. The higher factor should be adopted when conditions are likely to cause cycle times to fluctuate significantly.

The mean interval between the departures of cars when there are c lifts operating together is then

$$T_c = FT_1/c \qquad (2.3)$$

where F is an arbitrary factor between 1·05 and 1·10.

The corresponding flow rate is then

$$a_c = E_p/T_c \qquad (2.4)$$

The size of the lift cars may be taken as that of the smallest standard car which has a capacity 25 per cent or more above E_p. If, for instance, $E_p = 12·0$, a 1200 kg car with a nominal maximum capacity of sixteen would be selected; this would result in a probability of about 1 in 10 that a passenger would be unable to board the first lift to leave after his arrival in the lobby. Figure 2.2 shows the car sizes that are required to reduce the probability of the car being full. Given the mean number of passengers on the horizontal axis, the upper curve indicates on the vertical axis the limit to the actual number of passengers in 99 per cent of cases. The lower curve indicates the 90 per cent confidence limit.

The graph is based on the assumption of a Poisson distribution of the number of passengers, which can occur with lift departures at fairly even intervals. Significant irregularity in departure times results in greater variation in the number of people carried, with larger lift cars required if the same proportion of passengers is to enter the first car after arrival.

Extending Example 2.3 with calculations for two and three cars gives the results illustrated in Fig. 2.3. The calculation involved is especially suited to the use of a computer, and the program required is very short. If the tables of E_h to E_p are punched on cards or stored on a tape or disc file the relationship between the cycle time of any lift system and the passenger arrival rate can be obtained just by inserting as data into the computer the lift-system parameters. The total processing time is found in practice to be smaller when the unidirectional traffic tables are computed directly on each occasion (as described

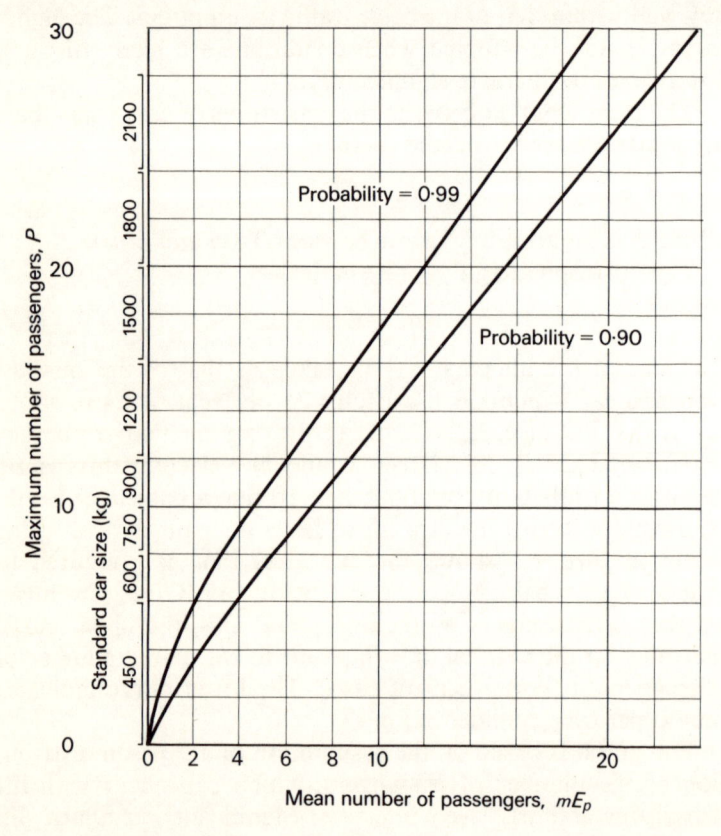

Fig. 2.2 Probability of there being P passengers or fewer wishing to enter a lift car
For unidirectional traffic, $m = 1$
For random inter-floor traffic, $m = (n + 1)/4n$, when n is odd, or $(n + 2)/4(n + 1)$ when n is even

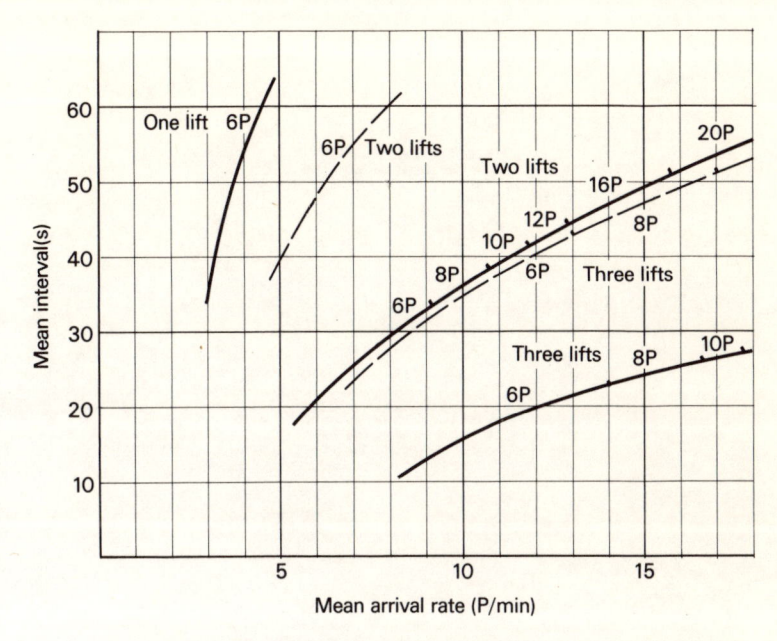

Fig. 2.3 Variation of inter-departure interval with arrival rate. 1·5 m/s lifts serving ground floor and 6 upper floors, 3·3 m storey height. Continuous curves: unidirectional traffic from the ground floor. Broken curves: random inter-floor traffic. Required lift car capacity (persons) given beside curves

later) and only the inter-floor values of E_h to E_p are stored than when both sets of tables are either transferred or computed, but the cost differences are small. A page of typical output is reproduced as Fig. 2.4.

When the height of the ground-floor storey is greater than the remainder, or if for any other reason there is a constant additional component of the round trip time, the value of T_1 should be increased by an appropriate value. The most important instance of this occurs when a group of lifts serves an upper zone and travels express through a number of floors, in which case the figure to be added to T_1 is equal to $2Ht_h$, where H is the number of non-stop storeys, or, alternatively, $2d/v$, where d is the additional distance travelled in one direction and v the speed of the lift.

28 The Design of Interior Circulation

```
FLOORS ABOVE GROUND:  9
    1.5 M/S LIFT, GEARED V-V MOTOR,
1.1 H DOORS PREMATURE OPENING, 3.3 M STOREY HEIGHT.
LIFT PARAMETERS   2.00   2.00   0.00   2.20   6.80   6.80   6.80   3.00

INTERFLOOR TRAFFIC
 P0    PROB.   ONE LIFT    TWO LIFTS   THREE LIFTS   FOUR LIFTS   AV CAR WT
        IDLE    A    T      A    T      A    T       A    T
0.650   0.00   8.4  277    16.0  145    24.0  97    32.0  73    10.8  20  0
0.700   0.00   7.5  256    14.3  135    21.5  90    28.6  67     8.0  20  0
0.750   0.00   6.6  234    12.6  124    18.8  82    25.1  62     7.6  16  1
0.800   0.00   6.0  215    10.8  112    16.1  75    22.5  56     5.4  12  0
0.820   0.00   5.2  205     9.2  108    15.3  72    22.0  54     4.8  11  1
0.840   0.00   4.5  194     8.7  102    13.7  68    17.7  51     4.8  11  0
0.860   0.00   4.0  183     7.0  96     12.5  64    15.3  48     2.8  11  1
0.880   0.00   4.0  172     6.9  90     10.7  60    13.2  45     2.4   6  2
0.900   0.00   3.8  158     6.5  83     9.7   55    12.2  42     4.8   6  1
0.910   0.01   3.2  151     5.6  79     9.3   53    12.0  40     1.4   6  1
0.920   0.01   3.2  142     5.0  74     8.3   50    10.0  37     1.8   6  0
0.930   0.02   2.8  134     4.2  70     7.1   47     0.8  35     3.6   6  0
0.940   0.04   2.5  124     4.2  65     7.1   43     5.5  32     1.0   6  0
0.950   0.06   2.2  110     3.8  58     6.4   39     8.5  29     0.5   6  0
0.960   0.09   2.2  100     3.7  52     6.5   35     7.3  26     0.0   6  0
0.970   0.11   1.8   84     3.7  44     5.5   30     0.0  22     0.0   6  0

FLOORS ABOVE GROUND:  10
    1.5 M/S LIFT, GEARED V-V MOTOR,
1.1 H DOORS PREMATURE OPENING, 3.3 M STOREY HEIGHT.
LIFT PARAMETERS   2.00   2.00   0.00   2.20   6.80   6.80   6.80   3.00
```

INTERFLOOR TRAFFIC

P0	PROB. IDLE	ONE LIFT A	ONE LIFT T	TWO LIFTS A	TWO LIFTS T	THREE LIFTS A	THREE LIFTS T	FOUR LIFTS A	FOUR LIFTS T	AV CAR	WT
0.700	0.00	8.0	296	15.2	155	22.7	103	30.3	78	10.7	0
0.750	0.00	7.0	271	13.3	142	20.0	95	26.7	71	8.6	1
0.800	0.00	6.5	247	11.3	130	17.8	86	22.7	65	6.7	0
0.820	0.00	5.5	236	10.6	124	15.8	83	21.1	62	5.9	1
0.840	0.00	5.1	224	10.8	118	14.7	78	19.0	59	5.2	0
0.860	0.00	4.7	212	9.0	111	13.4	74	17.2	56	4.5	1
0.880	0.00	4.3	190	8.1	104	12.2	70	16.2	52	3.8	0
0.900	0.003	3.0	183	7.2	96	10.0	64	14.5	48	3.2	1
0.910	0.003	3.0	176	6.8	92	10.25	61	13.6	46	2.8	0
0.920	0.003	3.0	167	6.3	87	9.5	58	12.6	44	2.5	1
0.930	0.01	3.0	155	5.4	81	8.3	54	10.8	41	2.2	1
0.940	0.02	2.5	144	5.0	75	8.1	50	10.8	38	2.0	0
0.950	0.03	2.5	131	4.4	69	7.3	46	8.8	34	1.5	0
0.960	0.07	2.2	117	4.4	61	6.6	41	8.8	31	1.2	0
0.970	0.13	2.1	90	3.9	52	5.9	35	7.8	26	0.9	0

A   MEAN ARRIVAL RATE, PASSENGERS/MINUTE
T   MEAN INTERVAL BETWEEN LIFT DEPARTURES, SECONDS
AV   MEAN NUMBER OF PASSENGERS
CAR   LIFT CAR SIZE
WT   PERCENTAGE OF PASSENGERS UNABLE TO ENTER CAR

Fig. 2.4 Output of a typical lift-calculation program

To aid calculation of lifts serving upper zones, graphs E1–E4 of Appendix 1 illustrate the values of T_1 when a number of local floors is served by standard cars loaded to about 80 per cent maximum capacity.

Example 2.4

A group of lifts is to serve floors 15–29 of a building of storey height 3·3 m. Select the number and size of lifts required to carry 20 P/min from the ground floor to this zone of upper floors.

The procedure is to calculate the mean interval and corresponding arrival rate of a number of alternative systems, selecting the cheapest that satisfies the criteria.

Selecting arbitrarily a speed of 2·5 m/s, the extra time per cycle taken in travelling fourteen extra storeys is $2 \times 14 \times 3·3/2·5 = 37$ s. Referring to graph E2, the cycle time for the fifteen local storeys served is, with a 1200 kg car, 133 s. The total cycle time is

$$T_1 = 133 + 37 = 170 \text{ s}$$

The corresponding arrival rate is, using eq. (2.2),

$$a_1 = (12·8 \times 60)/170 = 4·5 \text{ P/min}$$

(12·8 being the value of E_p given on graph E2). Hence $T_5 = 36$ s and $a_5 = 21·5$ P/min.

Other solutions are:

3·5 m/s lift, 1200 kg car.	$T_1 = 152$ s, $a_1 = 5·1$ P/min; $T_4 = 40$ s, $a_4 = 19·2$ P/min; $T_5 = 32$ s, $a_5 = 24·0$ P/min
3·5 m/s lift, 1500 kg car.	$T_1 = 168$ s, $a_1 = 5·7$ P/min; $T_4 = 44$ s, $a_4 = 21·8$ P/min
5·0 m/s lift, 1200 kg car.	$T_1 = 140$ s, $a_1 = 5·5$ P/min; $T_4 = 37$ s, $a_4 = 20·8$ P/min
5·0 m/s lift, 2100 kg car.	$T_1 = 170$ s, $a_1 = 6·8$ P/min; $T_3 = 60$ s, $a_3 = 19·4$ P/min.

Three large lifts with a contract speed of 5·0 m/s could carry a flow of passengers close to the given arrival rate of 20

P/min, but with a mean interval of 60 s. For half this interval, five lifts would be needed. As five cars of 1200 kg at 3·5 m/s could carry 24 P/min with an interval of 32 s, at the given arrival rate the actual interval would be below 30 s. Four cars, with speeds 3·5–5·0 m/s yield intervals of about 40 s. It will be seen that only by increasing the number of cars in a system can a substantial reduction be made from a calculated interval. An increase of car capacity gives a greater capacity but a correspondingly greater interval. An increase of speed yields a small increase of capacity and an improved interval; the greater the non-stop distance travelled, the greater the effect of car speed.

The capacity of a lift system for down-peak traffic may be checked with the use of Fig. 2.5. Assuming that the number of

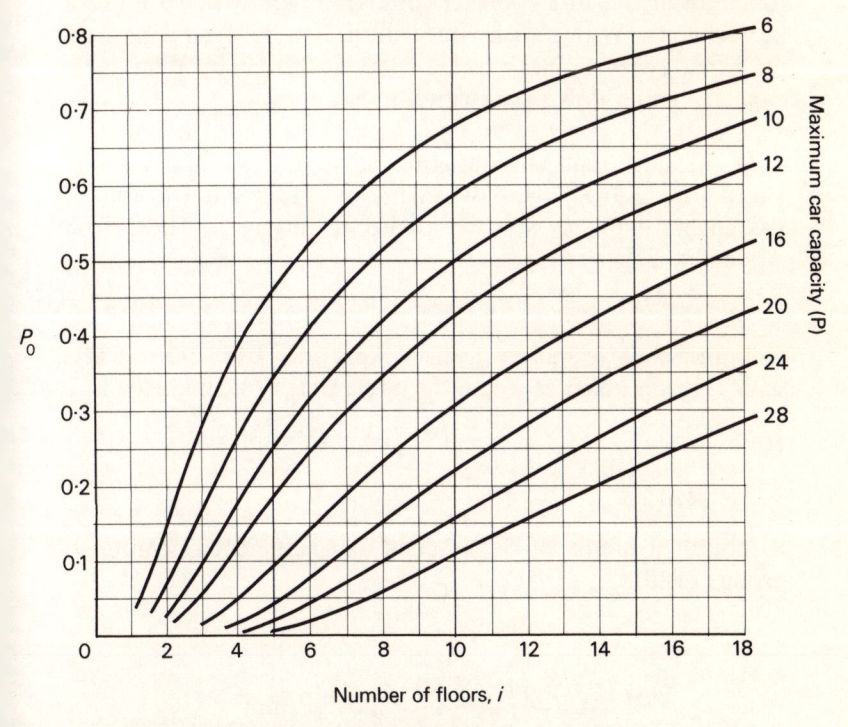

Fig. 2.5 Down peak traffic, uniform distribution
Traffic rate at which the probability of a lift car being full is 0·1 after being available at i floors

people wishing to enter lift cars matches a Poisson distribution, the graph indicates the traffic rate (in terms of P_0) at which the probability of the lift being full after being available at a given number of floors is 0·1. For instance, with down-peak traffic and 12 P (900 kg) lift cars serving all storeys of a building with 10 upper floors, the cars will be full before arriving at floor 2 on 1 occasion in 10 (on average) when the traffic flow is such that P_0 is about 0·35. This value may then be used in conjunction with the tables of Appendix 2 (unidirectional traffic) to calculate the cycle time and the actual passenger arrival rate.

Example 2.5

A building with thirteen storeys above ground is served by four lifts of 2·5 m/s contract speed and cars with 16 P (1200 kg) capacity. With a mean interval of 30 s between departures the capacity for up-peak traffic is about 22 P/min. What is the capacity under down-peak conditions?

(a) Assume that all four cars serve the 13 upper floors. From Fig. 2.5, $P_0 = 0.4$ when the probability is 0·9 that all passengers will be able to board at floor 1. Then, from Appendix 2,

$E_h = 12\cdot 3, E_{s1} = 4\cdot 9, E_{s2} = 1\cdot 8, E_{ss} = 2\cdot 1, E_p = 11\cdot 9$

Adopting the parameters given in Appendix 1 for 2·5 m/s lifts, with t_p being taken as 2·1 s, the cycle time of a single lift is

$T_1 = 12\cdot 3 \times 1\cdot 32 \times 2 + 4\cdot 9 \times 7\cdot 0 + 1\cdot 8 \times 6\cdot 9 + 2\cdot 1 \times 6\cdot 8$
$\quad + 11\cdot 9 \times 2\cdot 1$
$= 118\cdot 5 \text{ s}$

Adopting a factor of 5 per cent for inefficient operation of groups of lifts,

$$T_4 = \frac{118\cdot 5 \times 1\cdot 05}{4} = 31\cdot 1 \text{ s}$$

$$a_4 = \frac{11\cdot 9 \times 60}{31\cdot 1} = 23\cdot 0 \text{ P/min}$$

Thus with a mean interval of just over half a minute between successive lifts arriving at floor 0, some twenty-three passengers per minute can be carried. This is only, in fact, up-peak flow seen in reverse, a mean rate of 1·8 P/min per floor.

If it were to be assumed that virtually all passengers from floors 2 and 1 walked down, a different calculation could be made, treating the system as serving 11 upper floors with two express storeys. The resulting cycle time and arrival rate would be found almost the same as in the first case but, being distributed over 11 floors instead of 13, the effective traffic capacity would be about 2·1 P/min per floor.

(b) Assume that the lifts are divided into groups, two lifts serving only floors 9–13, the other pair serving only the lower storeys.

Upper zone (5 upper floors)
From Fig. 2.5, $P_0 = 0.09$. Adopt 0·10.

From Appendix 2,

$$E_h = 4\cdot9, E_{s1} = 4\cdot1, E_{s2} = 0\cdot3, E_{ss} = 1\cdot0, E_p = 11\cdot5$$

Eight additional storeys non-stop give an extra term, (1·32 × 8 × 2 s), in the expression for T_1. It is then found that $T_1 = 95\cdot8$ s, $T_2 = 50\cdot3$ s, $a_2 = 13\cdot7$ P/min. The mean interval, about 50 s, is now greater, but the effective capacity of 2·7 P/min per floor is also greater.

Lower zone (6 upper floors; assume that people on floors 1 and 2 usually walk down).
From Fig. 2·5, $P_0 = 0\cdot14$. Adopt 0·15.
From Appendix 2,

$$E_h = 5\cdot8, E_{s1} = 4\cdot5, E_{s2} = 0\cdot6, E_{ss} = 1\cdot1, E_p = 11\cdot4.$$

Two additional storeys non-stop, 1·32 × 2 × 2 s.

$$T_1 = 87\cdot7 \text{ s}, T_2 = 46\cdot0 \text{ s}, a_2 = 14\cdot9 \text{ P/min}$$

This is a capacity of 2·5 P/min per floor with a slightly smaller interval than in the upper zone. At the cost of longer waiting times the traffic capacity of the zoned systems is about 25 per cent greater than the comparable single system.

The curves in Fig. 2·3 showing lift performance under random inter-floor traffic were calculated from eqs. (2.1)–(2.4)

34   The Design of Interior Circulation

and the appropriate tables of Appendix 2. As this traffic pattern gives the minimum car occupancy, and as tolerance of additional waiting due to fully loaded cars appears to be less with inter-floor traffic than morning peak conditions, the car size selected should be at least those defined by the 0·01 probability curve in Fig. 2.2. It is these that are shown in Fig. 2.3.

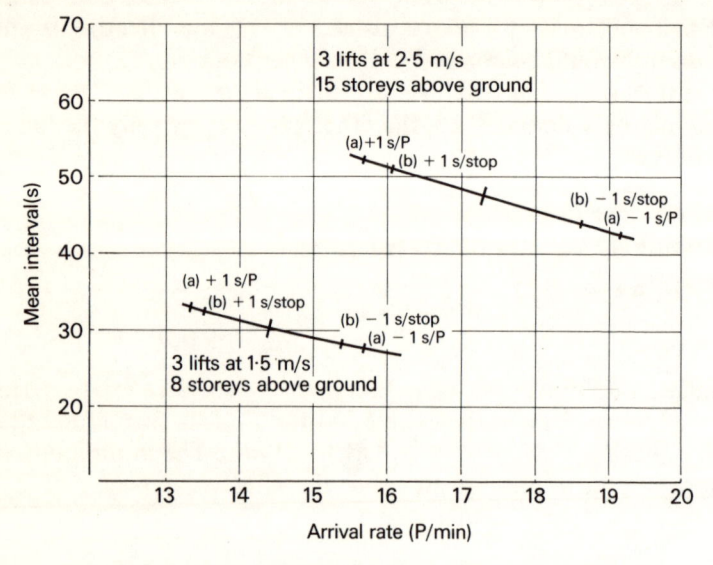

Fig. 2.6 Sensitivity of lift-capacity calculations. Effect of variation of (a) ± 1 s/P in passenger transfer time, (b) ± 1 s/stop in levelling and door operating time, in two different lift systems with unidirectional traffic

Lift performance calculations are, unfortunately, quite sensitive to small variations in the system parameters, particularly those associated with the time spent at landings, E_{s1} to E_p. The magnitude of the variation is illustrated with two examples in Fig. 2.6. Lifts intended to carry heavy traffic flows are designed to minimise the time required for levelling, door operation and acceleration to contract speed, and differences in the performance of comparable systems by different manufacturers are due to differences in these operations. The times taken by passengers to enter and leave lift cars are affected by the design of both lobbies and cars, in particular by the width of the door opening in relation to the car floor area, and also by the number of people entering at each stop. The values of t_p adopted in the examples given, 2·1 s with unidirectional traffic and 3·0 s with inter-floor, reflect that when a group of persons moves in or out of a car with no opposing traffic the mean time per person is significantly less than when a small number of transfers both from and to a lift is made at each landing.

The tables of Appendix 2 are calculated for the case that each lift car returns to the ground floor during every cycle. This is fairly common, and it is found also that the trip time calculated on this basis is a fair estimate of the cycle time of cars under a control system where some redundant journeys may be made to reduce variation in passenger waiting times at landings. If the control system is such that the assumption is inappropriate with random inter-floor traffic, the values of E_h and E_{s1} should be replaced by E'_h and E'_{s1} to give a good approximation of the shorter cycle time:

$$E'_h = 2E_h - n \qquad (2.5)$$
$$E'_{s1} = E_{s1} - p_0^{2n} \qquad (2.6)$$

Similarly when lifts are programmed always to travel the full height of the building, calling at the highest and lowest floors during each cycle, the values inserted into eq. 2.1 should be

$$E''_h = n \qquad (2.7)$$
$$E''_{s1} = E_{s1} + p_0^{2n} \quad \text{(inter-floor traffic)} \qquad (2.8)$$
$$E''_{s1} = E_{s1} + p_0 \quad \text{(unidirectional traffic)} \qquad (2.9)$$

2.3 Models of lift performance

The time that a lift takes to complete a round trip is dependent on the number of passengers boarding the car during the cycle. This is, in turn, partially a function of the time taken to complete the previous cycle – the longer the interval between departures from a landing the greater the probable number of people wanting to enter the car, at a given arrival rate. The performance of a lift can be seen as a sequence of events, some directly dependent on a random process, the arrival of potential passengers at landings, and some determined by earlier operations in the sequence which were themselves partly dependent on the distribution of arrivals.

The relationship between the cycle time of lifts and the rate at which intending passengers arrive can be examined by holding constant one of the factors and determining its interdependence with the other variables.

2.3.1 *Upward traffic, constant number of passengers in lift cars on departure*

We assume that:

(1) all traffic originates at the lowest floor, and the lift stops at upper floors only to allow passengers to leave the lift car;

(2) the number of passengers in the car as it departs from the ground floor is a value, p, which is the same on every journey;

(3) the intended destination of each passenger is independent of that of any other passenger and is equally likely to be any upper floor.

The expectation of the number of stops made by a lift during a round trip is then

$$E_s = n\left[1 - \left(\frac{n-1}{n}\right)^p\right] + 1 \qquad (2.10)$$

This equation was first published by Bassett Jones[79] and is given by Phillips[118] in an expanded form that allows the probability of destination to vary between the upper landings. An expression for the upper landing at which the lift reverses has been given independently by a number of authors:

$$E_h = n - \sum_{i=1}^{n-1} (i/n)^p \qquad (2.11)$$

where the floors are numbered from the ground upwards, 0, 1, 2, ..., i, ..., n. Substituting in a shortened form of eq. (2.1), this gives

$$T_1 = 2E_h t_h + E_s t_s + p t_p \qquad (2.12)$$

where t_s is the average time lost through braking, door operation, and acceleration at each stop.

Equations (2.2)–(2.4) can be used (with the substitution of p for E_p) to derive the corresponding passenger arrival rate with one or more lifts.

This analysis is simple and widely used. Despite the rigid assumption of a constant number of passengers in a lift car on departure, a condition unlikely to occur even when a substantial queue has formed, the equations given are the basis of most lift calculations in practice and are found to be satisfactory when applied with discretion to morning arrival peak traffic.

2.3.2 Upward traffic, constant probability of a call

Here we depend on a slightly more flexible assumption, that the probability remains constant of one or more passengers arriving during each cycle to travel to any given upper floor. As before, it is taken that the mean number of passengers intending to travel to each upper floor is the same.

Letting p_0 be the probability that during a cycle there will be nobody arriving to make the journey to any particular upper floor, the expectation of the number of stops made by a lift car during a round trip is

$$E_s = n(1 - p_0) + 1 \qquad (2.13)$$

Using combinatorial analysis this can be broken down to give the expectation of the number of stops preceded by a journey of a given height. If k represents a number of storeys ($1 \leqslant k \leqslant n$), the expectation of the number of stops during a cycle preceded by a run of k storeys is

$$E_{sk} = \sum_{m=1}^{n+1-k} p_0^{n-m}(1-p_0)^m \times \frac{m(n-k)!}{(m-1)!(n+1-k-m)!} + (1-p_0)p_0^{n-k} \qquad (2.14)$$

The two terms of the right-hand part of the equation represent the upward and downward sections of the round trip. It follows that

$$E_s = \sum_{k=1}^{n} E_{sk} + p_0^n \qquad (2.15)$$

where p_0^n is the probability of the lift becoming idle, of no passengers arriving during the period of the previous cycle.

The probability of the lift travelling to a given floor, i, but no higher is $(1-p_0)p_0^{n-i}$. The expectation of the highest floor is, therefore

$$E_h = \sum_{i=1}^{n} i(1-p_0)p_0^{n-i} = n - \sum_{i=1}^{n} p_0^i \qquad (2.16)$$

Alternatively,

$$2E_h = \sum_{k=1}^{n} kE_{sk} \qquad (2.17)$$

The equations for E_s and E_h, (2.13) and (2.16), become identical with those based on a constant number of passengers, (2.10) and (2.11), as n, the number of upper floors, increases towards infinity.

To determine the average number of people carried during each cycle it is necessary to define the probability distribution of which p_0 is one value. We will assume that passenger arrivals are independent and random, and that the probability of there being r passengers in the lift travelling to any given floor is

$$p_r = \frac{(\alpha)^r}{r!} e^{-\alpha} \qquad (2.18)$$

This is a Poisson distribution with a mean number of passengers α.

Then, with $E_p = n\alpha$ and $p_0 = e^{-\alpha}$,

$$E_p = -n \ln p_0 \qquad (2.19)$$

The mean cycle time and the corresponding round trip time can again be calculated with eqs. (2.1)–(2.4).

The probability of there being not more than P passengers in a lift car is

$$\text{prob}\,(p \leqslant P) = \sum_{p=0}^{P} \frac{(n\alpha)^p}{p!} e^{-n\alpha} \qquad (2.20)$$

Values of this expression may be obtained from standard tables of the cumulative Poisson distribution and are also illustrated in Fig. 2.2.

2.3.3. Random inter-floor traffic, constant probability of a call

With a lift serving the ground floor and n others there are $n(n + 1)$ different journeys that can be made between pairs of floors, counting upward and downward journeys separately. Because it is the case that causes a lift to make the greatest number of stops we shall assume that the traffic intensity is the same between every pair of floors; let α denote the mean number of people arriving to take the lift from one given floor to another during one cycle and p_0 the probability of no person arriving. We shall also assume, because it is a common case and causes a small increase in the cycle time, that the lift car travels to the ground floor during every cycle whether or not a call has been registered there. Then the car travels upwards to pick up all calls that have been registered for journeys to higher floors since the departure of the previous lift, reverses at the highest landing demanded, and visits in sequence all floors where a downward call has been registered. The period between successive reversals at the ground floor is taken to be the cycle time.

Taking, as before, a Poisson distribution of the number of people arriving during a cycle, the expectation of the number of passengers carried during a round trip is

$$E_p = -n(n + 1) \ln p_0 \tag{2.21}$$

The expectation of the highest floor reached can be derived as follows:

Probability of the lift travelling to floor $n = 1 - p_0^{2n}$
Probability of the lift travelling to floor $n - 1$ but not to floor $n = [1 - p_0^{2(n-1)}]p_0^{2n}$

Similarly, the probability of the lift travelling to a lower floor, i, but no further is

$$[1 - p_0^{2i}]p_0^{2n}p_0^{2(n-1)}p_0^{2(n-2)} \ldots p_0^{2(i+1)} = [1 - p_0^{2i}]p_0^{(n-1)(n+i+1)}$$

Thus

$$E_h = \sum_{i=1}^{n} i[1 - p_0^{2i}]p_0^{(n-1)(n+i+1)} \tag{2.22}$$

The expectation of the number of stops can be found by combinatorial analysis [146] but the resulting expression is exceedingly long and there are computational difficulties in evaluating it directly. The function is, however, tabulated in Appendix 2.

There appears to be no theoretical reason why expressions for the number of stops made and distance travelled should not be found for any pattern of traffic or any type of lift control program. The difficulties are practical and lie both in the tedium of deriving the equations and in the large numbers involved in the computation. The use of Monte Carlo methods gives another approach which is in most instances far more useful, and this is described in Section 2.3.7.

2.3.4 Variation in cycle time

Unless the cycle time of the lift is constant, the assumption of a Poisson distribution of passengers in a lift car is not the same as that of a Poisson distribution of people arriving to take the lift. Although few surveys have been made of the frequency distribution of arrivals in a lift lobby, a Poisson distribution has been found in other areas of pedestrian movement and it has been used extensively as a simple basis of queueing analysis.

Similarly, lift cycle times can be described by an Erlangian distribution:

$$b(t) = \frac{(\mu k)^k}{(k-1)!} e^{-\mu k t} t^{k-1} \qquad t > 0 \qquad (2.23)$$

This is a special case of the Gamma density function and is related closely to the Chi-squared distribution. The mean value is $1/\mu$ (corresponding with T_1 in eq. (2.1)), and k is a parameter that describes the degree of variation, so that $k = \text{mean}^2/\text{variance}$. When $k = 1$, t varies as the period between successive arrivals in a Poisson distribution; as k approaches infinity, t becomes constant. The Erlangian distribution can be viewed as the superimposition of k Poisson processes and at moderately high values of k it differs little from a Normal distribution.

If, with a Poisson distribution of arrivals of mean a, the probability of r people arriving in a period t is

$$\pi_r(t) = \frac{(at)^r}{r!} e^{-at}$$

then the probability of r people arriving during a cycle with an Erlangian distribution is

$$p_r = \int_0^\infty \pi_r(t) b(t) \, dt = \frac{(r + k - 1)!}{(k - 1)! r!} \frac{(\mu k)^k a^r}{(\mu k + a)^{k+r}} \quad (2.24)$$

This is a form of the function known as the Negative Binomial Distribution. It has been tabulated by Williamson.[164]

Thus, from (2.24),

$$p_0 = \left(\frac{\mu k}{\mu k + a} \right)^k \quad (2.25)$$

The expectation of r is a/μ, mean arrival rate multiplied by mean cycle time. Therefore, with n upper floors and unidirectional traffic from the ground floor,

$$E_p = kn(1/p_0^{1/k} - 1) \quad (2.26)$$

and with random inter-floor traffic

$$E_p = kn(n + 1)(1/p_0^{1/k} - 1) \quad (2.27)$$

It can be shown that these approach eqs. (2.19) and (2.21) as k approaches infinity. At lesser values of k (that is, with some variation in lift cycle times) E_p is larger than when cycle time is constant, increasing, in turn, the mean cycle time. This implies that if, with a queue of people waiting in a lift lobby, the period between departures begins to vary, the average number of people in each car increases and so also does the mean interval. However, as the division of E_p by the cycle time gives the relationship between the number of people carried and the mean interval, the differences tend to cancel and the graph of cycle time against passenger flow rate changes little. The numerical difference between values of E_p calculated with constant and Erlangian distributions of cycle times is very small for all but the lowest values of k, so in practice the assumption of a constant cycle time is a satisfactory approximation.

In the same way it is possible to find the distribution of passenger waiting times under moderately busy conditions. If the arrival time of a person at the lift lobby is independent of the position of a lift in its cycle, and the passenger never fails to board the car when it arrives, then the probability of his having to wait longer than a given time T is

$$W(T) = \frac{1}{k} e^{-\mu k t} \sum_{j=1}^{k} \frac{j(\mu k T)^{(k-j)}}{(k-j)!} \qquad T > 0 \qquad (2.28)$$

with an Erlangian distribution of cycle times.

The expectation of the waiting time is

$$E_w = (k + 1)/2\mu k \qquad (2.29)$$

When $k = 1$, a Poisson distribution of cycle times,

$$E_w = 1/\mu$$

the mean waiting time is equal to the mean interval.

For a constant cycle time

$$E_w = 1/2\mu$$

The nearer the cycle time is to being constant, the less the average wait. Using eq. (2.24) to find the probability of the number of arrivals exceeding a given car size it is possible to calculate the number of passengers waiting after the first car has departed, but extending this to consider passengers waiting after two or more departures involves a compounding of probabilities and becomes complex. Bailey[11] has discussed this problem in the context of patients waiting for clinic appointments.

The variability of cycle times decreases as the rate of passenger arrivals is increased, largely because the number of stops made during a round trip approaches a maximum. At a given arrival rate both the mean and the variance of cycle times become greater when the number of floors served is increased; for a constant number of storeys the mean cycle time decreases and the variance increases as the pattern of traffic is changed from a uniform distribution to one in which some floors generate a greater passenger flow than others.

Gaver and Powell[60] have derived an expression for the variation of round-trip time with up-peak traffic. The results they have obtained, for lifts serving upper zones of a fifty-storey building, correspond with values of k of 296 and above in an Erlangian distribution. In other surveys values around $k = 10$ have been found with random inter-floor traffic, a much greater coefficient of variation, and this would be a conservative figure to adopt in calculations when data from comparable lift systems are unavailable.

2.3.5 Bunching of lift cars

When two or more lifts are linked under directional collective control the cars tend to bunch together when the traffic flow is heavy. If one lift passes through part of a cycle more slowly than normal (being delayed at a landing or answering a cluster of calls at adjacent floors) the distance separating the car from the one following it is reduced. The second lift is then likely to have fewer calls to answer and travel more quickly. The state soon occurs in which several lifts will call in quick

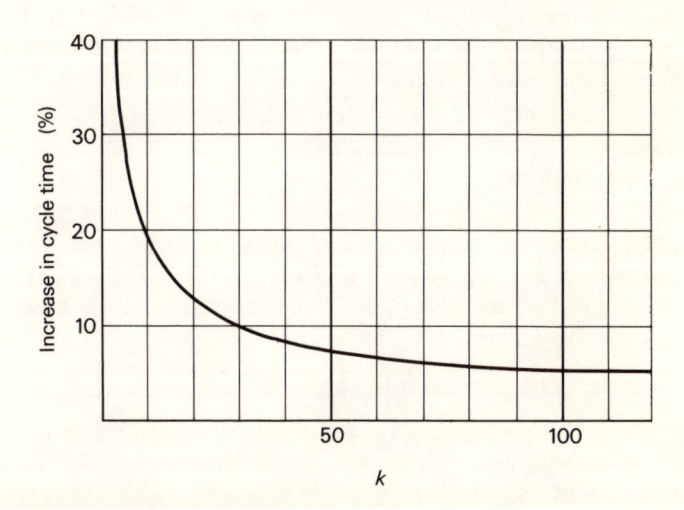

Fig. 2.7 Effect on cycle times of imposed delays to lift cars

succession at landings, to be followed by a long interval. There is a close similarity with the observed behaviour of buses during a rush hour.

It can be prevented by imposing a delay at one or more points in the cycle so that a lift car cannot leave a landing until a certain time has elapsed since the departure of the previous car; the optimum imposed interval is a fraction $1/c$ of the cycle time of a single lift carrying the same traffic as each of the c cars in a combined system, although this is difficult to effect when the rate of passenger arrivals varies. The mean cycle time is slightly increased by this imposed delay. Normally this is assumed to be covered by the arbitrary increase of 5–10 per cent in the calculated value, but if the variance of the natural cycle time can be estimated, the increase – the expectation of the actual delay – can be found from Fig. 2.7. The graph is based on an integration of the difference between two Erlangian distributions.

2.3.6 Doors re-opening

It is possible, with heavy inter-floor traffic, for a car to be delayed several times during a cycle by people who arrive just as car doors are closing, causing them to re-open. The practice is undesirable, and it tends to occur less in a satisfactory lift system than when the inter-departure interval is long, but if necessary an estimate of the mean delay can be found by the following method.

Let q denote the probability that one or more passengers will arrive just as the car doors are closing. This may occur more than once during each stop, and the expectation of the time by which the lift is delayed is

$$t_r[q(1-q) + 2q^2(1-q) + 3q^3(1-q) + \ldots]$$

where t_r is the time taken by the doors re-opening and beginning to close again.

If q is sufficiently small for the series to converge rapidly (that is, if the probability of the lift car becoming full is negligible), then the expression may be shortened by summing to infinity to give the approximate value.

$$E_{re} = t_r \frac{q}{1-q}$$

With Poisson arrivals of mean rate a at a particular floor, and a door closing time t_c,

$$q = 1 - e^{-at_c}$$

Thus

$$E_{re} = t_r(e^{at_c} - 1) \qquad (2.30)$$

2.3.7 Digital simulation

The Monte Carlo method is a technique of evaluating functions by a sampling process, using sequences of random numbers (or, more usually in computation, of pseudo-random numbers) and obtaining approximate solutions. In analysis of lift performance it can be used both to obtain numerical solutions of equations defining E_h, E_{sk} and E_p, and to create a detailed simulation of lift-system operation.

As an example of the first case, the values of E_h and E_{sk} for random inter-floor traffic may be found in the following way:

(1) Let each element of a two-dimensional array, **A**, represent the occurrence or otherwise of a journey by one or more passengers to or from a given floor in a given direction during one cycle of the lift car; the array will have $2(n + 1)$ elements for a lift system serving $n + 1$ floors. Let an element take the value 1 if the car is required to stop at the floor for passengers to enter or leave, and 0 otherwise. Let p_0 represent the probability that no passenger will travel from one given floor to another during one cycle. There are $n(n + 1)$ different journeys possible.

(2) Generate a sequence of $n(n + 1)$ random numbers ($0 \leqslant R \leqslant 1$). Let each number correspond with one possible journey. Set the value to 1 of any element of **A** that represents the origin or the destination floor of a journey to which the

corresponding random number is greater than p_0. Let the value of every other element be 0.

(3) On completion the array is an image of the calls made throughout the system during one cycle. Count the number of storeys separating the highest and lowest floors that the lift is required to visit and the number of floors visited that are preceded by non-stop flights of $1, 2, \ldots, n$ storeys.

(4) Repeat steps 1–3 a large number of times, with different sequences of random numbers. Calculate the average distance travelled by the lift per cycle and the mean values of the number of stops preceded by different flights.

Values of p_0 that differ from one journey to another and alternative rules by which the lift, or lifts, are assumed to visit the array of floors can be adopted to model uneven traffic patterns and different lift-zoning tactics.

Example 2.6

The building has four storeys above ground level. There is uniform inter-floor traffic with a probability of 0·85 that during any one cycle no passenger will travel from one given floor to another.

The first two columns in Table 2.4 give the origins and destinations of all possible journeys. The third column lists twenty random numbers (obtained from a table of random digits) allocated in sequence to the respective journeys. Array **A** shows the resulting pattern of calls for the lift.

During this cycle the lift is required to travel the full height of the building, stopping at all floors on the upward part of the trip and at floor 1 on the return. Successive repetitions, with an analysis of journeys based on the assumption that the car returns to the ground floor at the end of each trip, would lead to mean values of approximately 3·5 for the highest floor reached, 3·5 for the number of stops per cycle preceded by a one-storey journey, 0·9 for the number of stops preceded by two-storey flight, and 0·6 for the number of stops preceded by three or more storeys. If the lift is assumed not to return automatically to floor 0 when no call has been recorded there, analysis of journey lengths must take into account the lowest floor reached during the previous cycle.

Table 2.4 Simulation of lift-call patterns

Floor of origin	Destination floor	Random number	$p_0 = 0.85$ random number greater than p_0?
0	1	0.66	
0	2	0.94	yes
0	3	0.57	
0	4	0.11	
1	0	0.00	
1	2	0.51	
1	3	0.86	yes
1	4	0.88	yes
2	0	0.23	
2	1	0.65	
2	3	0.67	
2	4	0.43	
3	0	0.24	
3	1	0.71	
3	2	0.05	
3	4	0.06	
4	0	0.17	
4	1	0.89	yes
4	2	0.61	
4	3	0.74	

		Passenger journey: Upward	Downward
Array **A**	Floor 4	1	1
	3	1	0
	2	1	0
	1	1	1
	0	1	0

Although two or three repetitions can be made manually in a fairly short time a computer is necessary if the estimates of travel distance and stopping frequency are to lie within narrow confidence limits. The number of repetitions required is related to the variance of the results; the more the cycle time of the lift would vary, the greater the degree of uncertainty in the estimate with a given number of repetitions. The true value of the mean or the variance of the quantities estimated, the value that would emerge from an infinite number of repetitions, will lie to some degree above or below the actual value obtained, and the range within which it will lie with a given level of probability is inversely proportional to the square root of number of repetitions made. The figures of E_h and E_s for random inter-floor traffic listed in Appendix 2 were calculated this way. To obtain a probability of 0·999 that the values obtained of E_{s1} (the quantity showing the greatest variation) were within $\pm 0·1$ of the true number, 400 repetitions were necessary for a four-storey system and 1800 repetitions for eighteen storeys. The variance is greater at low traffic intensities (high values of p_0) than with very busy systems.

This sampling procedure is used just for the evaluation of a function relating the number of stops made by a lift with the number of floors served and the probability of calls occurring. The other application of the Monte Carlo method is in forming a model of continuous lift operation. Here it is taken that time elapses with successive repetitions of the programme; during each cycle the incoming traffic is determined, the response of the lift system calculated, and the resulting waiting and journey times found. A record is kept, and revised in each cycle, of the calculated number of passengers in various parts of the system and the location and movement of cars, statistics which become the output of the procedure. The time scale of a simulation need not be regular, with a fixed period between the beginnings of successive cycles, and for modelling a process, such as the operation of a lift system, in which critical events occur at short but irregular intervals, the time base of the calculation is more conveniently advanced in step with these events. As a lift car approaches a landing, there is a point at which the control system must begin a series of operations if the lift is to stop, a call made at the landing or in the car being answered only if made before this point is reached. The progress of cars between braking points can be represented by successive cycles of a simulation, and the outline of a simulation program with this basis is shown in Fig. 2.8.

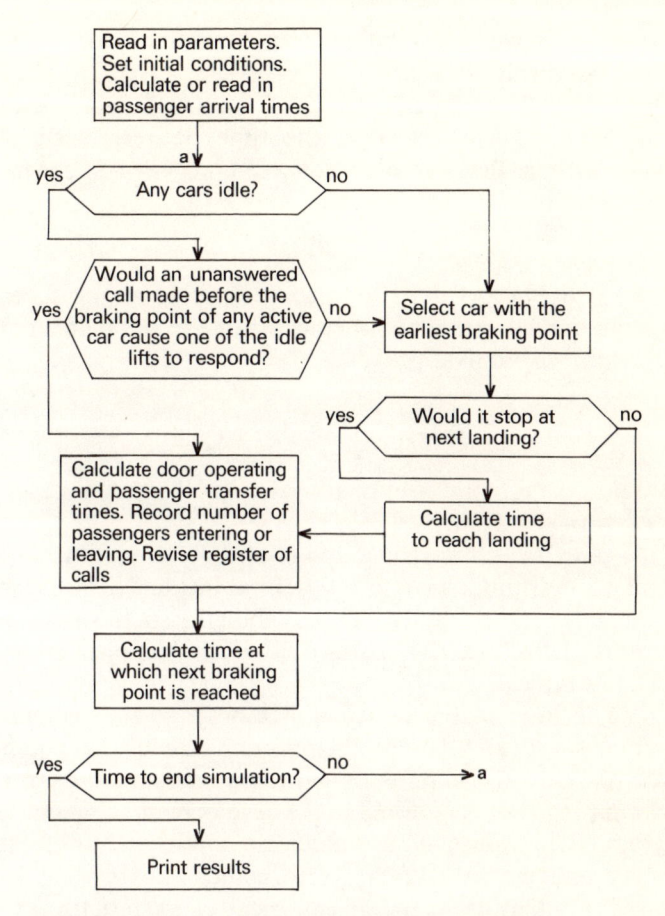

```
┌─────────────────────────┐
│ Read in parameters.     │
│ Set initial conditions. │
│ Calculate or read in    │
│ passenger arrival times │
└─────────────────────────┘
            │ a
            ▼
yes    ╱ Any cars idle? ╲  no
   ◄──╱                   ╲──►
       ╲                 ╱
        ╲               ╱
         ╲             ╱
            │
            ▼
      ╱ Would an unanswered ╲
      ╱ call made before the ╲      ┌──────────────────────┐
 yes ╱ braking point of any active ╲ no │ Select car with the  │
◄───╱  car cause one of the idle   ╲──►│ earliest braking point│
     ╲         lifts to respond?  ╱    └──────────────────────┘
      ╲                         ╱                 │
       ╲                       ╱                  ▼
                                         yes ╱ Would it stop at ╲ no
                                        ◄───╱   next landing?    ╲───►
                                             ╲                   ╱
                                                     │
                                                     ▼
┌──────────────────────────┐          ┌──────────────────┐
│ Calculate door operating │          │ Calculate time   │
│ and passenger transfer   │◄─────────│ to reach landing │
│ times. Record number of  │          └──────────────────┘
│ passengers entering or   │
│ leaving. Revise register │
│ of calls                 │
└──────────────────────────┘
            │
            ▼
┌──────────────────────────┐
│ Calculate time at        │
│ which next braking       │
│ point is reached         │
└──────────────────────────┘
            │
            ▼
yes   ╱ Time to end simulation? ╲ no
  ◄──╱                           ╲────► a
      ╲                         ╱
            │
            ▼
     ┌──────────────┐
     │ Print results │
     └──────────────┘
```

Fig. 2.8 Simplified flow diagram of a lift-simulation program

The central cycle of the program begins at point 'a'. Passenger arrival times may be calculated as required during the simulation, or determined initially for the whole simulation period, as shown. They may be read as initial data (being, perhaps, actual values from a survey) but in most instances will be calculated using random numbers scaled according to a given function, such as a Poisson distribution. Methods of obtaining a specific frequency distribution from a sequence of random numbers, by sampling or by the use of inverse functions, are well covered in texts on simulation.[142] Two useful relationships are these:

Given that R is one of a series of random numbers evenly distributed between 0 and 1, the time between successive arrivals with a Poisson distribution, a negative exponential distribution, is

$$t_i = -a \ln R \qquad (2.31)$$

where a is the mean number of people arriving in unit time.

For an arrival rate that varies with time in the manner of a typical morning arrival peak,

$$a = \frac{A}{T} \frac{15}{4} \left[\frac{2}{15} + \left(\frac{t}{T}\right)^2 - \left(\frac{t}{T}\right)^4 \right] \qquad (2.32)$$

where A is the total number of arrivals, T is the total time and a is the instantaneous arrival rate at time t.

This gives an asymmetric graph of arrival rate against time, a reaching a maximum when $t/T = 0.71$ at just under three times the rate when $t/T = 0$ and 1. The combination of the above two equations yields an arrival pattern of which the upper graph of Fig. 2.1 is typical.

There is little mathematical difficulty in writing computer programs for lift simulation. The programs can be complex and therefore expensive to prepare and to operate, and a degree of programming skill is required in the development of an efficient program. The values of a simulation model are, first, that virtually any measure of lift performance can be calculated and, second, that arbitrary or empirical characteristics of lift performance and passenger flow can be incorporated.

It is possible to treat this form of mathematical model as a piece of experimental apparatus, measuring the results of controlled changes in the input variables and in the structure of the system just as physical laboratory models can be used in other fields of investigation.

The results from simulating traffic patterns must be treated with no lesser degree of statistical caution than numbers obtained from surveying a real system. A model of lift performance during a single morning arrival peak can give no more information than observation of the traffic within an actual building on just one morning. The statistical design of the experiment should be based on the same principles as survey work; in addition it is necessary to show that the assumed patterns of pedestrian movement and lift-system operation do not differ, in their relevant characteristics, from those in the building which the model is intended to represent.

Several descriptions of lift simulation by computer have been published; some simulation programs are available commercially.[13]

2.3.8 Choice of model

There is little point in making calculations to a fine degree of precision when either the anticipated traffic flow rates or the criteria that a lift system is designed to satisfy are known only as coarse estimates. Frequently neither are known well. It is general practice, then, to introduce into computations some factor to compensate for a possible underestimate. This may well have the effect that the majority of lift systems can carry heavier flows than actually occur, but occasional severe failures do arise from misjudgment of likely traffic patterns. Changes from one computational model to another will hardly affect the matter. More efficient design requires a more complete analysis of accommodation within a particular building and further research generally into the factors that determine passenger satisfaction.

Given a fair estimate of the morning peak traffic, a flow upwards, that is, from the ground floor of the building, simple calculations based on the assumption of a constant probability of a call are adequate in most cases (the model for which graphs are given in Appendix 1 and tables in Appendix 2). It is unlikely that the population of the building will actually be distributed evenly over all floors or that the arrival of potential passengers will match exactly a Poisson distribution, but the effect of divergence from these is, in all but the most extraordinary instances, a slightly better inter-departure interval than calculated; the magnitude of error is small in comparison with that resulting from moderate changes in door operating, levelling or passenger movement times. The less detailed model based on the assumption of a constant number of passengers in the car on departure will serve equally well when a substantial factor of safety has been added to compensate for uncertainty in the estimate of traffic flow, but as this involves hardly any less calculation than the more accurate case (given that tables of E_h and the other constants are available) the advantage is negligible.

To deal with inter-floor traffic is harder. Not only may the pattern of traffic vary substantially within a building during a working day but the precise response of the lifts depends upon the type of control system. Unfortunately among the operating systems there is none best on all counts: some, for instance, that allow relatively short waiting times under heavy traffic are less efficient with intermittent flows. In addition, an estimate of inter-floor traffic is usually more difficult to obtain and usually even less precise than those of the unidirectional flows as the building fills and empties. Often the best that can be done is to calculate the inter-floor traffic that a given lift system could carry under the most demanding conditions and this is the best use of the analytical model given. As before, the assumption that traffic is random and uniformly distributed gives, almost always, the longest calculated cycle times. The model can be usefully extended in the ways described when greater precision is warranted.

If the traffic flow is both accurately known and complex in its distribution, especially if the effect of its variation with time is to be studied, Monte Carlo methods are valuable. When a simulation model can be formulated so also can a model based on analysis of alternative probabilities, but the latter may become so complex that evaluation is entirely impracticable. The very ease of obtaining results by simulation can, though, be misleading, for the completeness and apparent realism of the output may obscure the fact that initial data are hypothetical. It is good practice in the presentation of simulation results to give limits of confidence based on analysis of the sensitivity of the output to changes in input traffic flows and lift characteristics.

2.4 Capacity of paternoster lifts

2.4.1 Capacity limits

Paternoster cars move continuously, at a constant speed, and analysis of the traffic flow is less complex than that of conventional lifts. Cars travelling in each direction pass landings at intervals of about 10 s, and the capacity of the system is limited entirely by the size of the cars. In a standard paternoster system each car can hold two people.

Although a potential passenger has only a short wait at a landing before a car arrives, he is not necessarily able to board it. One or two passengers may be already in the car and there is time for only two transfer movements – for two people to either enter or leave the car, or for one transfer in each direction. The resulting relationship between the number of passengers in a car on its arrival at a landing and the number in it on departure is given in Table 2.5. A similar table can be formed to give the number of people entering a car on each occasion.

The total rate at which passengers can enter cars depends on the pattern of traffic flow. With people moving in one direction from a single floor – upward traffic from the ground floor of a building, for instance – the maximum rate of traffic flow is $2/T$, where T is the interval between successive departures of a car, the storey height divided by car speed. By contrast, with pas-

Table 2.5. Number of passengers in a paternoster car on its departure from a landing

Number in car on arrival	Number leaving car	Number of people waiting at landing: 0	1	2 or more
0	0	0	1	2
1	0	1	2	2
	1	0	1	1
2	0	2	2	2
	1	1	2	2
	2	0	0	0

sengers queueing at all floors of a building and making short regular journeys of one or two storeys, it is possible to devise traffic patterns where the total flow is $2(n + 1)/T$ passengers in unit time (the system serving $n + 1$ floors). This is the ultimate capacity of a paternoster but, unfortunately, such cases are unlikely to occur in real conditions.

2.4.2 Random inter-floor traffic

Assume that passengers move independently, that the floor to which a passenger travels is not determined by the destination of any other passenger. Let L denote the probability that a randomly selected passenger in a paternoster car will leave at the next landing. There may be 0, 1 or 2 people in the car; the probability of 0, 1 or 2 leaving has a Binomial distribution:

$$\text{prob } (k \text{ passengers leaving}) = \binom{m}{k} L^k (1 - L)^{m-k}$$

where m is the number of passengers in the car and

$$\binom{m}{k} = 0 \qquad k > m$$

$$\frac{m!}{k!(m-k)!} \qquad \text{otherwise}$$

where $0!$ is taken as 1.

Let C_m be the probability of m passengers in the car arriving at the landing, C'_m the probability of m on departure (and hence on arrival at the next landing) and Q_0 and Q_1 the probabilities of 0 and 1 person waiting to board the lift. Then, using Table 2.5, the transition probabilities are

$$C'_0 = C_0 Q_0 + C_1 L Q_1 + C_2 L^2 \qquad (2.33)$$

$$C'_1 = C_0 Q_1 + C_1(1 - L)Q_0 + C_1 L(1 - Q_0)$$
$$\qquad + C_2 2L(1 - L)Q_0 \qquad (2.34)$$

$$C'_2 = C_0(1 - Q_0 - Q_1) + C_1(1 - L)(1 - Q_0)$$
$$\qquad + C_2(1 - L)^2 + C_2 2L(1 - L)(1 - Q_0) \qquad (2.35)$$

If Q_j represents the probability of j people waiting at a landing at the start of one cycle and Q'_j the probability of there being j people at the start of the next, we have

$$Q'_j = q_j E_0 + q_{j+1} E_1 + q_{j+2} E_2 \qquad j \geqslant 1 \qquad (2.36)$$
$$Q'_0 = q_0 + q_1(E_1 + E_2) + q_2 E_2 \qquad (2.37)$$

where $E_0 = C_2[L^2 + (1-L)^2]$
$E_1 = C_1 + C_2 2L(1-L)$
$E_2 = C_0$
$$q_j = \sum_{r=0}^{j} Q_{j-r} P_r$$

and $P_r =$ the probability of r potential passengers arriving during one cycle.

If there is a persistent queue of two or more passengers at every floor the actual number of people waiting does not affect the number being carried by the lift, and the system can be considered a finite Markov chain. The number of passengers entering thus approaches statistical equilibrium even though the lengths of queues may not. Under these saturation conditions C_1 must be zero, and the expectation of the number of passengers entering at a given landing is

$$E_e = C_2 2L(1-L) + 2C_0 \qquad (2.38)$$

When a car is always empty on arriving at the lowest floor at the start of the upward journey then, at this floor, $E_e = 2$. At the next floor $C_0 = 0$ and $C_2 = 1$, and eqs. (2.33) and (2.35) will give the values C_0 and C_2 for the higher floors in succession. For a model of the growth of queues under moderate flows of traffic, we make the further assumptions:

(1) The mean rate of passenger movement is the same between each floor and every other. Then, if the car is travelling upwards approaching floor i ($1 \leqslant i \leqslant n$), the probability that a given person will leave the car is

$$L = \frac{1}{n+1-i} \qquad (2.39)$$

(2) The patterns of people arriving at landings to take the lift may be described by Poisson distributions: the probability of r

passengers arriving during any cycle to travel upwards from floor i is

$$P_r = \frac{A^r}{r!} e^{-A} \tag{2.40}$$

where $A = (n - i)\, ad/v$
and $\quad a =$ the mean number of passengers travelling from one floor to another in unit time
$d =$ the storey height
$v =$ the speed at which cars travel (e.g. $0 \cdot 30$ m/s).

Equations similar to (2.39) and (2.40) can be given for downward traffic.

(3) When the system starts operating, there are no queues and every car is empty: for all floors $Q_0 = 1$ and $C_0 = 1$.

No passengers are carried at the start of the first cycle; in the next cycle the number of people in each car leaving a landing is equal to the number of arrivals there during the first cycle, provided that this was not more than two people. In the third cycle, the passengers already in the car start to affect the rate of entry. If the arrival rate at every floor is less than the maximum possible rate of entry there, it is found numerically that the probabilities of given queue sizes approach constancy.

The state of a system after a period of operation may be found from successive evaluations of eqs. (2.31)–(2.35). It may also be determined by Monte Carlo simulation; longer computing time is necessary than with direct use of the transition-probability equations but the amount of storage required is much less.

When queues have formed at every floor with random interfloor traffic, the rate at which passengers are carried is shown by curve (a) of Fig. 2.9. But the mean number who can enter a car is not the same at each level. If two passengers enter every car at the lowest floor, at the next landing a person can enter only if just one of the original pair leaves there. The mean rate of boarding is lowest at floor 1 (for upward traffic) and it increases with successive storeys.

Under lighter flows, and Poisson arrivals, the curves generated are (b) and (c) of Fig. 2.9. Queues can build up fairly slowly under an unchanging distribution of arrivals, but as

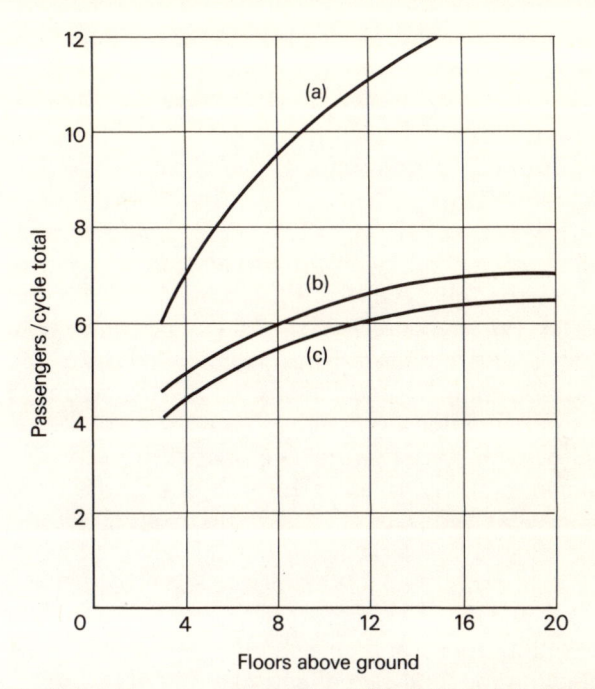

Lifts and Escalators: Traffic-flow Calculations 57

Fig. 2.9 Variation of paternoster capacity with building height. Random inter-floor traffic
(a) Total capacity with queues at every floor
(b) Poisson arrivals: traffic intensity at which the mean rate of entry at one or more floors is just equal to the arrival rate after 100 cycles operation
(c) Poisson arrivals: traffic intensity at which the system is close to statistical equilibrium after 100 cycles (Cycle time = (height between cars)/speed)

queues develop so does the average number of passengers in each car, and the rate at which people can enter declines. After starting with the system empty of passengers the rates of flow that cause saturation after $100T$ s are given by the upper of the two curves. It shows the levels of traffic at which the rate of passengers entering becomes equal to the mean rate of arrivals at one or more floors. Curve (c) gives the arrival rates that lead to statistical stability, where the mean number of passengers in the system rapidly approaches a maximum. These rates represent a reasonable design capacity for continuous random traffic and correspond with a probability of about 0·6 that at the busiest floor no passenger will be left queueing after the departure of a car.

When traffic is concentrated between only a few floors of a building the overall capacity is less than with movement between every pair of landings. The curve should be read at the point that corresponds with the actual number of levels used rather than the full height of the system. For short periods a paternoster can take much higher flows of traffic than those that result ultimately in statistical equilibrium, up to the maximum rate of entry at one landing of $2/T$ passengers in unit time in each direction of travel.

It has been observed in practice that when queueing occurs at landings (and it is usually the result of brief heavy flows of people, such as a discharging audience from a lecture room) some passengers will walk up or down stairs a storey to enter the cars at an earlier floor. With many people waiting at ground level it is found also that some tend to enter downward-moving cars, travelling round and up again to by-pass the queue. It has little effect on the total rate of entry.

Example 2.7

A paternoster serves the ground floor and seven upper floors of a building. The storey height is 3·5 m and cars are spaced at one-storey intervals. If the car speed is 0·3 m/s,

$$T = 3·5/0·3 = 11·7 \text{ s}$$

With traffic randomly distributed between all floors the system can carry 5·4 P/cycle (Fig. 2.9) or $5·4 \times 60/11·7 = 27·8$ P/min. For traffic between fewer levels the capacity is less; distributed over four floors, the capacity (for prolonged traffic flow) is 4·1 P/cycle (Fig. 2.9, reading for three floors above ground) which is equivalent to 21·0 P/min. The maximum rate at which the paternoster can carry passengers in one direction from a single landing is $2 \times 60/11·7 = 10·3$ P/min.

If the occupants of the building are accustomed to using a paternoster an increase of speed to 0·4 m/s would give capacities 33 per cent greater, but with users unfamiliar with the equipment, hesitation in boarding faster-moving cars might cause a decrease in the effective rates of passenger movement.

2.5 Capacity of escalators

The capacity of an escalator is given by the rate at which passengers can step on to the moving stairs. Except as it affects this rate, the behaviour of people actually travelling on the escalator, running up the steps for instance, is unimportant.

The entry rate is found to vary with the following factors:

(1) Width between balustrades.
(2) Operating speed. At low speeds a limit is determined by the available standing room on steps. As speed is increased the number of empty steps becomes greater; the rate of entry increases, reaches a maximum, and then declines as passengers are deterred by the apparent speed and step with caution. Observations of a number of escalators in London Underground Stations[149] showed that with 1·2 m wide escalators a maximum of 175 P/min occurred at 0·74 m/s (measured up the slope). This is illustrated in Fig. 2.10.
(3) Queueing at the entrance to the escalator. It has been observed that, with crowding, people tend to step on to the escalator alongside one another instead of occupying a whole step as at lesser traffic densities. The increase in capacity is lessened by a greater boarding time per person. Fruin[58] has recorded that passengers boarding a 1·2 m escalator took about 20 per cent longer under heavy traffic than at a low density.
(4) Passenger characteristics. Shoppers with bulky parcels, adults and children together, and elderly people tend to step more reluctantly than members of the commuting public. Escalators in department stores, therefore, carry lower rates of flow than those in transport terminals. In the survey by Fruin it was found that mean entry speed of men was about 10 per cent greater than that of women, with light traffic, and the effect of baggage was to reduce the average speed by 5 per cent. Con-

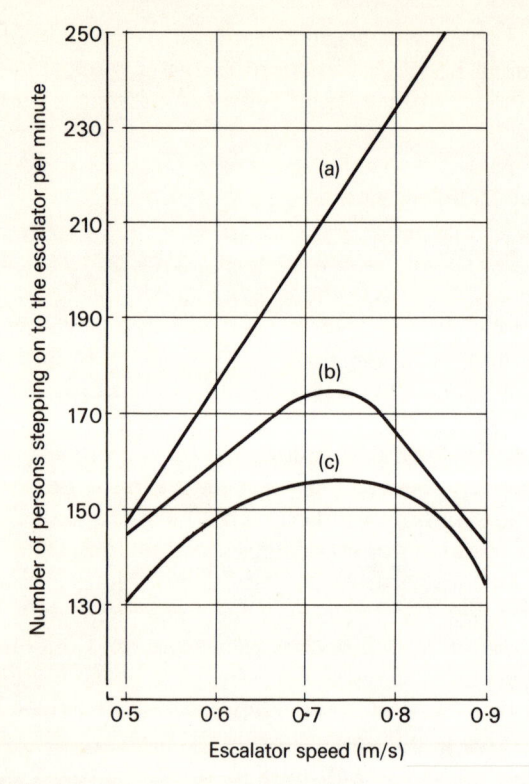

Fig. 2.10 Variation of escalator capacity with speed. Upward traffic in London Underground stations on escalators 1·2 m wide between balustrades
(a) Theoretical capacity with two persons per step
(b) Maximum number of persons stepping on to an escalator during 1 min
(c) Average rate during 5 min period

siderable variation was found in each category, the distribution of boarding times being close to Normal, a standard deviation of 0·25 s being found when the mean was about 1·0 s.

Initial design values for escalators are listed in Table 2.6. Other than the surveys referred to, there is little published research on escalator capacity. The upper limit of traffic flow can be found by multiplying the escalator speed, in steps per minute, by the number of people who could stand on each step, but this figure is reached only at very low speeds. Various authors recommend between 75 per cent and 80 per cent of the maximum as the normal design capacity, and the lower figure is given in Table 2.6. It will be seen, on comparison with Table 4.1, that the capacity of an escalator is approximately that of a corridor with the same overall width.

Table 2.6. Approximate capacity of escalators (P/min)[138, 149]

speed	1·2 m between balustrades			0·8 m between balustrades	
	limit (2·0 P/step)	0·75 limit	LTE survey	limit (1·25 P/step average)	0·75 limit
0·46 m/s (90 ft/min)	136	102		85	64
0·61 m/s (120 ft/min)	181	136	150	113	85
0·74 m/s (145 ft/min)			155		

In buildings such as department stores, and in pedestrian subways in shopping districts, the effective capacities of escalators can be half those given in the table. The proportion of the very young and the elderly among the users is significant and, because people are reluctant to step alongside strangers, the difference between wide and narrow escalators can be small. In the absence of detail a useful assumption for traffic calculations is 60 P/min per escalator, regardless of dimensions.

There must be sufficient clear floor area at the head of an upward-moving escalator and at the base of one that is downward moving to ensure that, even if a queue has formed at the next stage of the circulation route, passengers are not hindered in stepping off. Similarly, if escalators are designed to carry surges of pedestrian flow from the arrival of trains, for example, queueing space must be provided in the approach to the escalators. The design of waiting areas is described in Section 4.1.4.

Research on the traffic capacity of passenger conveyors or moving ramps is still required. Capacity is determined, as with escalators, by the rate at which passengers can step on; the operating speed, except as it affects this, is immaterial. As passenger conveyors are used in long circulation routes in airports and other transit buildings, a fairly high speed is desirable. Under BS 2655[20] this may be greater than the permitted speed of escalators, up to 0.90 m/s, a moderate walking speed. The entry rate at this speed may be slightly less than with a slower conveyor, if passenger behaviour is analogous to that on escalators. On the other hand it is possible that people are prepared to step rather more readily onto a level or only slightly inclined platform than onto one at the top or base of a staircase and, on balance, it might be reasonable to assume that the maximum capacity of a passenger conveyor 1.2 m wide between balustrades is the same as that of a similar escalator, approximately 155 P/min. At low speeds it can be assumed that passengers will stand as close together as in a stationary queue, about 0.5 m apart, and the rate at which passengers are carried past a point can be found by multiplying the operating speed by the average number of passengers standing within a unit length of the conveyor.

CHAPTER
3
Mechanical equipment

3·1 Traction and electro-hydraulic lifts

3.1.1 *Overall dimensions*

The majority of lifts have the form illustrated in Fig. 3.1(a), an electric traction machine set directly above a vertical shaft. Within the shaft are the car and counterweight, the guide rails on which they run, the landing door mechanisms, cables and control switches and, at the foot of the shaft, buffers.

The plan area of each shaft is approximately twice the internal area of the car in the majority of standard lifts, and about three times the floor area of very small cars. The vertical dimensions above and below the terminal landings vary with the size and speed of the lift; those of a 2·5 m/s 1200 kg lift are shown in the illustration. It will be seen that the distance from the highest floor served to the ceiling of the machine room is in this example nearly three times the storey height, and the depth of the pit at the base of the shaft is approximately equal to one storey.

The clear height required in the shaft above the highest floor served is related to the stroke of the counterweight buffer and

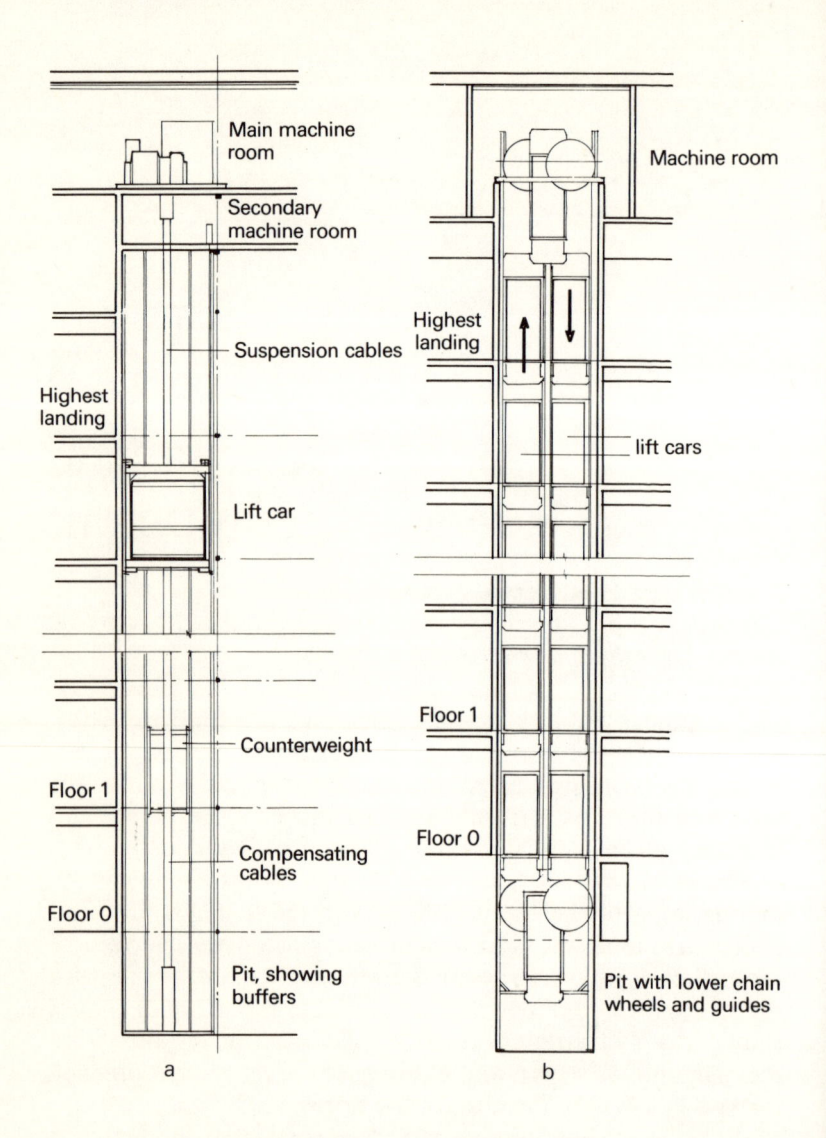

Fig. 3.1 (a) Overall view of an electric traction lift
(b) Overall view of a paternoster system

the clearance between the normal terminal position of the weight and the position at which the buffer is engaged. Should a car continue moving upwards above the terminal floor, and the normal limit controls fail to halt it, the counterweight would become supported by the buffer and traction lost at the driving sheave. The necessary car top clearance varies from about 2 m, with lifts of 1·0 m/s and below, to about 4 m with a contract speed of 5·0 m/s.

It is feasible, with slow lifts, to reduce the overall height of the system by placing the traction machinery lower in the building, at the side of the shaft, with cables taken over pulleys at the top. This, however, increases the initial cost; it is mechanically inefficient, and the saving in height small.

Buffers are required in the pit for both the car and the counterweight. They may be spring buffers if the contract speed of the lift is not greater than 1·0 m/s, but with greater speeds are self-resetting oil buffers. The pit may also contain a sheave, sliding in a short vertical track, to guide compensating cables which are suspended from the underside of the car to the counterweight to reduce variation in balance with changes of car position. With very fast lifts the sheave may be anchored to prevent the car or the counterweight jumping upwards when the downward motion of the other is arrested sharply; additional car top clearance is necessary when this is not provided. A pit approximately 1·5 m deep is required with a car speed of 0·5 m/s; this increases to approximately 5 m with 5·0 m/s.

With electro-hydraulic lifts driven by a direct-acting ram, the car top clearance need be only 600 mm or 75 mm above the projection of any equipment above the car roof, whichever is the greater. No overhead equipment is necessary as the motor room is usually located at ground level beside the shaft, and the roof of the lift shaft need often be no higher than that of the adjacent upper storey. Pit depth may also be less than in a traction system but the design of the base varies with the type of hydraulic system. The overall dimensions of standard electric lifts and the car sizes normally associated with particular contract speeds are given in BS 2655:Part 3.[20] Lifts of standard dimensions are considerably cheaper than those purpose-made.

3.1.2 Machine-room equipment

The loads and speeds of passenger lifts are not high in comparison with the demands on electrical machinery in some other uses, but the need for precise control of acceleration and braking, and the small levelling tolerances required, limit the choice of traction machinery.

Car speeds of up to 0·5 m/s may be provided with a single speed a.c. motor driving a worm reduction gear. The speed actually attained by a lift and the rate of acceleration depend on the slip of the induction motor, and this varies with the load. A variation of 30 per cent in maximum speed can occur between an empty car and one fully loaded. A spring-operated electromagnetic drum brake is an integral part of the traction machine and, as in other types of lift machine, is spring applied and electrically held open; the lift is stopped by switching off power to the motor and allowing the brake to close. Accuracy of levelling at a landing is therefore a function of the load, the speed and direction of travel, and the rate at which the brake is applied. An accuracy of ±25 mm can be achieved with severe braking, but this dimension is doubled if the brake is applied more slowly to give greater comfort to the passengers. Such a traction system is feasible for light pedestrian traffic but unsatisfactory when the lift is used for wheelchairs or trolleys. For occasional traffic this type of motor geared to give a car speed of 0·25 m/s gives smaller inaccuracies in levelling.

Greater comfort for the passengers, and better levelling, can be achieved with the use of a two-speed motor, the lower speed of which is engaged as a car approaches a landing stop. CP 407[23] recommends a ratio of 3:1 between normal running speeds of 0·75 and 1·0 m/s and levelling speed, the levelling accuracy then being ±20–25 mm. Cars with a maximum speed of 0·75 m/s may have a less expensive type of safety equipment than the faster lifts and are usually better value. For running speeds of 1·5 m/s and above a d.c. motor in a Ward–Leonard system is used for traction. The mains power drives an a.c. motor which turns a d.c. generator to provide current for the traction machine. Control of the generator field allows a variable voltage to be applied across the armature of the traction

Mechanical Equipment 67

motor, and this permits car speed to be changed smoothly. Slowing down is effected by a progressive reduction in motor speed to a very small levelling movement; the brake is used only to hold a stationary lift. Above 1·5 m/s contract speed (full running speed) the traction motors are gearless, the main sheave being mounted directly on the motor shaft. Gearless motors are heavier and more expensive but mechanical efficiency is greater, noise less, and the life of the machine significantly longer. The levelling accuracy of a variable-voltage lift should be ±10 mm. Figure 3.2 shows a typical installation of d.c.-geared machines.

Automatic levelling control on the faster lifts, and those with long travel, is maintained with transducers between the car and landing openings. The stretch of traction cables can cause a car to move more than the distance of tolerance when a load is transferred and a variable-voltage machine system allows a corrective movement that is imperceptible to passengers.

The mechanical characteristics of small lifts allow only a coarse control on the rates of acceleration and breaking. With high maximum car speeds and variable-voltage equipment the tolerance of passengers to changes of movement becomes important, because not only can a specified curve of acceleration be attained accurately but the overall performance of the lift is closely dependent on the additional journey time involved in landing calls.

Little systematic research on the acceptability of different rates of acceleration has been published. Menon[98] states that acceleration must be limited to 2·4 m/s^2 and the rate of change of acceleration to 1·2 m/s^2 if discomfort is to be minimised; Oplinger,[113] on the other hand, says that high values of the rate of change of acceleration are tolerable for short periods and values of 1·5–3·0 m/s^3-s ('jerk-seconds') appear to be satisfactory. This limits acceleration itself to the range of these numbers, which also contains the value given by Menon.

In practice the graph of velocity with time which describes the way a fast lift gains speed is neither exactly parabolic (a constant rate of change of acceleration) nor linear (a constant acceleration). But in the absence of specific information an approximation to flight times of lifts with maximum speeds

68   The Design of Interior Circulation

Fig. 3.2 A lift machine room with geared variable-voltage equipment. Two motor generator sets can be seen in the central area. There are two traction machines at the far side of the room and another can be seen in the foreground. The rectilinear units alongside are floor selector devices

1·5–5·0 m/s can be obtained by integrating on the assumption that the critical factor is the rate of change, with a value between 0·9 and 1·2 m/s³, lifts of 2·5 m/s and above having a value at the top of this range. The equations are as follows:

Assume that acceleration of a car increases, after leaving a landing, by a rate of change J to a maximum acceleration f_1. The rate of change then becomes $-J$ until at distance d_2 and time t_2 the acceleration is momentarily zero and the velocity is at a maximum, v_2. The car then begins to slow down, and the pattern of braking is the same, in reverse, as that of acceleration. Maximum braking occurs at time t_3; at time t_4 and distance d_4 the car is again at rest. Then

$$t_4 = 4(d_4/2J)^{1/3} \qquad (3.1)$$

The time and distance required to reach a speed v_2 (with zero acceleration at this point) are

$$t_2 = 2(v_2/J)^{1/2} \qquad (3.2)$$
$$d_2 = J(v_2/J)^{3/2} \qquad (3.3)$$

By substituting the maximum speed of the lift for v_2, the distance required to reach this can be calculated. Similarly,

$$t_1 = f_1/J \qquad (3.4)$$
$$d_1 = \tfrac{1}{6}J(f_1/J)^3 \qquad (3.5)$$

If a limiting acceleration value is specified, by substituting the value for f_1, the corresponding time and distance are given.

Within the machine room there is, for each lift, a traction machine mounted directly above the well, a motor-generator set (in variable voltage systems), control equipment, and a governor for operating safety equipment. The control equipment may be in two or more parts; in particular a floor selector device may be a separate unit and, with groups of fast lifts, a supervisory control unit may be separate. In the faster systems, also, the governor and a deflecting sheave may be accommodated in a secondary machine room within the shaft walls below the main motor room.

The plan area of the motor room is from three to five times

the well area, the ratio varying with car size and the type of equipment. For a specific installation the required motor-room dimensions may be found in BS 2655:Part 3 for standard lifts or obtained directly from the manufacturing company. The room height should not be less than 2·1 m and in most cases at least 2·6 m is necessary. A steel beam over the machinery should be provided to allow a hoist to be used during installation and maintenance. This may be aligned with a series of removable floor sections through the building for the initial delivery of the machinery.

Electro-hydraulic lifts are available for travel distances up to 20 m and speeds up to 1·0 m/s. This type of equipment is illustrated in Fig. 3.3. The car is moved by a direct-acting ram from a cylinder set within a borehole, being driven by oil pumped at a pressure between 1700 kN/m^2 and 7000 kN/m^2. Alternative types employ rams mounted within the lift shaft and acting either telescopically or through a roping system. With these the cost of the borehole is eliminated at the expense of increased mechanical complexity.

Car motion of hydraulic lifts is smoother and levelling more accurate than that of traction lifts with a.c. motors, alignment at landings with a tolerance of ± 6 mm being possible. The motor room is usually beside the base of the lift shaft but need not be immediately adjacent. It contains an oil storage tank, the motor and pump, and the valve equipment. Safety gear within the shaft is not required (in Britain) on lifts with a direct-acting ram when the cylinder can be withdrawn from the borehole for inspection, but an anti-creep device is required to prevent a car moving more than 75 mm at landings with changes in oil pressure. Motor room noise must be taken into consideration in the design of the building.

Larger motors are required for electro-hydraulic lifts than traction lifts as no counterweight is installed. The power consumption is thus greater, but not in proportion to the motor power ratings because the hydraulic lift motor operates only for upward travel. The total capital cost of the lift equipment and associated building work is marginally higher in most cases than that of traction lifts but maintenance costs are markedly lower.

Mechanical Equipment 71

Fig. 3.3 An electro-hydraulic lift. The photograph looks up the shaft to the underside of the car. The direct-acting ram is in the centre and the guide rails may be seen at the sides of the shaft. A landing door-operating mechanism is visible on the lower right-hand wall

3.1.3 *The car assembly*

The lift car is built within a rectangular steel frame to which the traction ropes are attached. On the underside of the frame is safety gear that grips the guide rails if the car speed becomes excessive; guide shoes are fixed to the top and bottom of the frame on each side to align the car during normal running. Figure 3.4 illustrates a small standard lift car.

The guide rails for car and counterweight are steel and of T-section. They are attached to the building structure by brackets, normally at one-storey intervals, which are bolted to steelwork or fixed to metal inserts in concrete. The cross-sectional dimensions of the rails are determined by the distance apart of the fixings and the weight of car or counterweight. To avoid unnecessarily large rails, when storey heights are greater than about 4 m, fixings should be provided more frequently. In the American Standard Safety Code for Elevators,[6] the required size of rail is tabulated against spacing distance and car or counterweight load; BS 2655 requires that deflection under normal operation does not exceed 3 mm. On gearless passenger lifts roller guide shoes are used, the three rollers on each shoe running on the machined sides and end of the T-section leg. Sliding shoes are cheaper and have some other advantages with slower lifts; for speeds above 1·0 m/s they should be self-aligning with automatic lubrication.

The safety gear on a car operates when the downward speed exceeds 115 per cent of the normal full running speed, the maximum tripping speed being specified in BS 2655 and other codes. On lifts with speeds greater than 0·7 m/s the gear is actuated by a steel or phosphor bronze rope passing from the safety gear to a governor pulley at the head of the shaft and a tension pulley at the base. If the car speed becomes excessive the governor operates first by cutting electrical power from the traction machine and brake; if the speed continues to increase the rope is gripped, the resulting tension causing the car safety mechanism to operate. Pairs of jaws become clamped on the guide rails, bringing the car to a halt. On cars with rated speeds of 0·75 m/s and below the action can be almost instantaneous, but for other lifts the mechanism must be such that the rails are

Mechanical Equipment 73

Fig. 3.4 A standard lift car in the factory. This is an eight-person car as used in housing projects. The door-operating mechanism and mechanical safety edge can be seen

gripped progressively. With very slow lifts a governor is not essential; the safety device may be operated by a rope which passes over an idling pulley to the counterweight and which is normally slack, failure of the main ropes causing it to be tensioned. In all cases it must be possible to release the safety gear by raising the car.

The passengers within the car must be enclosed, unable to touch the shaft walls or any mechanical parts. Ventilation is given by baffle openings, often just above floor level and at ceiling height so that air movement is induced as the car moves in the shaft, or by a small electric fan.

The weight of a car affects the power required for a specified acceleration pattern, the lower the weight the more economical the traction machinery. Walls and ceilings are usually, therefore, lightweight steel frame and panel construction with a steel, wood or laminated plastics finish, although almost any non-combustible surface is feasible. A problem associated with thin panel construction is resonant vibration, and noise is reduced by fixing to the main frame through absorbent pads and by ensuring that panels have a certain stiffness. BS 2655 requires that the car enclosure is not permanently deformed in any part by a moderate thrust over a small area. The lift floor is of heavier frame construction, designed so that eccentric loading causes negligible distortion, with sufficient strength to avoid failure under accelerated loading from the car halting on the buffers or on the operation of safety gear.

The appearance of the interior is affected markedly by the type of lighting. Such a small enclosure would be shunned by some of the users under other circumstances, and an appearance of 'lightness' is important. This is achieved principally by ensuring that the ceiling is high in reflectance and illuminance. A trans-illuminated ceiling is also satisfactory provided it is complete and not a central panel in an otherwise dark surface. The apparent size of the car is greater, in most cases, with walls of high reflectance than of dark materials.

Only passenger lifts intended for occasional use have manually operated doors. Efficiency is increased by the use of automatic opening and closing devices and these, in general, operate car and landing doors simultaneously, the two sets of doors being coupled and sliding horizontally. The closing kinetic energy must be limited as people or goods may occupy the doorway when the operation begins; the force required to prevent closing must also be limited so doors can be restrained manually. Three types of safety device are generally available: interrupted beam, lever-operated safety edge, and proximity-operated safety edge. In the first, one or more lamps are directed to shine on corresponding photoelectric cells on the other side of the door opening, an interruption of the light beam by a person or object causing the doors initially to retract or to be delayed in closing. Safety edges mounted on the leading faces of the car doors cause the doors to retract when an object presses the mechanical safety edge (which usually has a soft rubber surface) or is brought within a short distance from the electronic proximity device. The doors may be controlled so that, if an object continues to block the doorway, after a delay of some seconds the doors begin to close slowly and 'nudge' the obstruction. A bell or buzzer may sound as the doors close, but this can be a distracting noise elsewhere in the building.

The limit on kinetic energy is effectively a limit on door-closing speed. In large cars the doors are centre-opening, two leaves sliding apart, to give half the closing time of a single panel. The kinetic energy to be absorbed by an obstruction may also be reduced by forming each half into two quarter-panels which slide alongside one another.

The door control is interlocked with the traction machinery so that the lift cannot start until the doors are fully closed. In fast lifts the doors are controlled to open in the final moments of approach to a landing, when the car floor is within 250 mm of the landing floor, so that the leaves are fully apart by the time the car is stationary. Toe guards extend below the car and landing openings.

The indicators and controls required within the car are as follows:

(1) Direction indicator.
(2) Floor position indicator.
(3) Destination selection buttons. These may, by mechanical

or electrical means, indicate whether a particular landing has already been selected.

(4) Door opening button, to cause the doors to re-open or to hold them open. In the latter case, continuous pressure should be required, and the doors close normally once the control is released.

(5) Emergency communication. If the lift stops between landings it is necessary for the occupants to summon help and for those outside to talk to the people in the car. Except in buildings with general public access a telephone, connected to a caretaker's office or anywhere else occupied at all times the lift is used, is satisfactory; another telephone is fitted in the building beside the lifts and, often, in the machine room. Alternatively a bell push within the car may operate an alarm signal outside. The provisions necessary depend on the type of building and the size of the lift system.

For fast lifts each car should be fitted with a panel within the roof which opens outwards clear of any equipment and which, when opened, operates a switch to prevent the lift moving. In some cases this should be openable only from the outside. A control panel by which the lift can be moved slowly during maintenance is also installed on the car roof. Several safety precautions must be associated with these controls.

Vandalism or, at least, misuse is unavoidable in some forms of public construction, the isolation of a person within a lift car permitting the sort of defacement and destruction associated also with toilet areas; in lobby areas the incidence of wilful damage is less when exposed to supervision than when secluded. Store cupboards, meter cabinets and similar fittings are especially prone to damage when screened from view. In those housing projects where vandalism is likely, loitering spaces should be avoided by reducing to a minimum the area of internal circulation; the secondary use of lobby and corridor areas, for service and storage, is undesirable. Graffiti and fouling are discouraged by the use of rigid wall materials with strong texture and pattern. For lift cars ribbed aluminium sheet as a wall finish has been found suitable. A floor finish of PVC with integral coved skirting allows easy cleaning. Plastic control buttons and indicator numerals are prone to damage by prising out and by burning; metal buttons and non-projecting indicator devices should be fitted in public buildings. Light fittings should be flush moulded or concealed beyond reach.[157]

In all buildings, any use of passenger lifts for moving goods and furniture should be anticipated; this may affect the choice of internal finishes. For infrequent carriage of goods, passenger lift cars can be supplied with mats to protect wall surfaces.

3.1.4 Suspension

Although for slow lifts with short travel distances steel plate link chains may be used, the normal means of suspension is steel wire ropes which have an impregnated fibre core. A minimum static factor of safety of 10 is adopted, greater for fast lifts, and at least two independent ropes must be used.

In the simplest roping arrangement the cables run from the lift car to the driving sheave above the shaft and down to the counterweight assembly. The sheave is required to have a diameter at least forty times that of the rope, to control the rate of rope flexion, but except with small lifts the desirable sheave diameter is smaller than the distance from the centre of the car to the counterweight. A diverting pulley is therefore introduced to increase the distance separating the ropes and is mounted at the head of the shaft slightly lower than the traction sheave. This reduces the fraction of sheave circumference in contact with the ropes, and the resulting area may be insufficient to avoid slipping under high stress, even if the grooves in the sheave are V-shaped. For fast lifts, then, a double-wrap roping arrangement is used, the ropes from the car passing over the traction sheave, around the secondary pulley, over the traction sheave again, and across the pulley to the counterweight position. The substantial increase in area of contact between the rope and the sheave allows the locating grooves to be U-shaped, which gives longer rope life.

An arrangement in which the ropes are secured at the head of the lift shaft and pass under pulleys from which the car and counterweight are suspended, returning over the drive sheave and diverting pulley, allows the speed of sheave rotation to be doubled, and thus the motor may be smaller. This system is found in heavy freight lifts and some passenger lifts up to 2·5 m/s.

The lift counterweight is constructed of sectional metal weights within a rectangular steel frame. It is guided by rails similar in shape to the car guide rails, though not necessarily of the same cross-sectional area, and there are four guide shoes on the counterweight frame. The total weight is usually that of the car plus 40 per cent of the maximum car load. Counterweights should be fitted with safety gear if there are basements beneath the lift shaft, or any other spaces to which access is possible. It is also necessary, in such a case, to ensure that the lift pit floor or other buffer supports have sufficient strength to withstand impact of the moving car or counterweight.

When at the extremities of travel the imbalance of car and counterweight, due to the differing rope lengths to each, can be significant. For travel distances over 30 m compensating cables are usually necessary, and are suspended (as illustrated in Fig. 3.1) between the underside of car and counterweight around a guide pulley in the pit. On slow speed lifts, chains can be used instead of the steel wire ropes for this purpose.

3.1.5 *Operating systems*

Under all but the lightest traffic flows a lift that responds to calls in the order in which they are made, taking one passenger to his destination landing before commencing the journey for the next, travels much unnecessary distance. A lift control system is required, therefore, to maintain a register of calls and determine a sequence of landing stops that gives the maximum traffic-handling capacity or minimises some measure of passenger waiting times.

The majority of control systems are based on 'directional collective' operation. Two control buttoms ('up' and 'down') are fitted at each intermediate landing and lift cars have a set of buttons that enables a passenger to register his destination floor. As a lift car travels up the building it calls, in sequence, at all landings at which the 'up' button has been pressed and all for which a call has been recorded from the car. After reaching the highest floor demanded by upward-travelling passengers, the lift answers 'down' calls in a similar manner. As a car departs from a floor the record of the call is cancelled and following cars will not stop until the appropriate button is pressed again. If no further calls have been registered on the arrival of a lift at a landing, the car may remain there, standing with the doors open, or return to park at the ground floor. With two or three cars operating together in one system, one car may remain at the last landing served and the remainder park at ground level. In such systems, known as 'duplex' and 'triplex', an idle lift will re-commence operation when calls occur behind a moving car, and under various other conditions (which differ, to some extent, between systems produced by different manufacturers).

In multi-storey car parks and in residential buildings where all journeys have the ground floor as origin or destination, a 'down collective' system can be appropriate. Only downward calls can be made from upper landings, and the car stops in sequence to answer these; on upward trips the lift stops only at floors registered within the car.

With single lifts in small buildings (up to four storeys) simpler automatic operation is possible. One landing button is provided at each landing to bring the car to that level. An indicator light is illuminated when the car is unavailable. One call at a time is served, with uninterrupted journeys, but clearly the traffic capacity is very low; this system should not be used when the probability is significant of a potential passenger arriving to find the lift busy, generally with more than 5–10 calls per hour. In some buildings – hotels and department stores, for instance – lift attendants are employed. The use of a lever-operated switch by which the operator controls directly the movement of a car is obsolescent, the push-button control with automatic acceleration, braking and levelling having superseded this, and a standard control system is usually provided where an operator can over-ride landing calls with a key-operated switch. An additional indicator may be fitted within the car to inform the attendant of landing calls. The 'car-preference' switch that allows an operator to over-ride the normal control system can be required where hospital stretchers are carried and in other buildings where individual service is needed.

Under heavy traffic, especially with three or more lifts, directional collective control systems have a number of disadvantages. In particular:

(1) Cars tend to bunch together. Instead of car departures from a floor being evenly spaced in time, several cars may serve the floor within a short period, followed by a long interval.

(2) Under downward traffic flows, such as those that occur as office workers leave for home, lifts become filled at the upper storeys and people waiting lower in the building are unable to board. A load-weighing device may prevent full cars stopping, which cuts out unnecessary delays from the cycle time, but the disparity of service between upper and lower levels remains unchanged.

(3) The response to a surge in traffic flow is slow; queues develop while cars remain idle.

The first two of these may be overcome by ensuring that there is a certain delay between departures of successive cars from landings (an analysis of which is discussed in section 2.3.5) and by arranging for one or more cars, under heavy downward traffic, to return empty to an intermediate floor rather than to the highest landing at which a call is registered.

Control devices that are sensitive to changes in car loading, to the delay between the registration and answering of calls, and to the frequency at which calls are recorded are the basis of 'group supervisory' control systems. The response of the lifts varies under these, with the intensity and direction of traffic flow. There may be four or more different modes of operation: up and down peak flows, two-way traffic, intermittent traffic, and variants of each of these. Cars are taken out of service when the rate of traffic flow declines and become operative in a particular mode as an increased rate is detected. In most instances the modes of operation are those of directional collective control modified with automatic timing and allocation of cars to reduce the variations in waiting times found with collective control in its basic form. With logical switching devices of some complexity, using computer components, the lifts can be made to respond sensitively to variations in passenger demand and to react appropriately when contingencies, such as a pronounced delay to the departure of a car, occur.

The collective-control algorithm is not the only one that is feasible and, particularly under varying inter-floor traffic, other procedures can give better service to waiting passengers. In addition, the allocation of cars to calls can be improved, reducing stopping frequency and travel distances, if instead of a simple 'up' or 'down' demand the destination floor is registered at the landing from which a call originates. Such systems have been described by Port,[120] Closs[32] and by Barney.[13] Levy[91] has described another use of on-line computer control of lifts, applicable to high buildings in which the upper floors are grouped into different zones.

The analyses of lift performance in the previous chapter were based on assumptions of car movement under inter-floor traffic that do not match those occurring with certain of the more advanced control systems; the concept of an identifiable 'cycle time' is difficult to maintain. Given the specification of a system there is no reason (except, perhaps, very complicated combinatorial analysis) why the distributions of passenger waiting times could not be calculated; in most cases Monte Carlo methods will give the quickest results. However, on the assumption that the more complex control systems give rather better lift performance than those on which the standard calculations are based, these calculations will yield the lower limit of flow capacity, which is usually adequate in practice.

3.1.6 *Building construction work*

The lift well is an integral part of the structural framework of a tall building. The wall thickness is usually determined by the ratio of the building's core breadth to height; in concrete this is generally at least 125 mm and in a very high building may be much greater. Slip-form construction is frequently used. A steel-framed building may have a braced steel shaft with infill panels of brick or blockwork, but the use of concrete lift towers in a building that is otherwise steel-framed is common. Lift wells in small buildings are usually formed in brick or block work. Except when lifts are intended for use in firefighting it is not necessary to have a separate containing shaft for each car; a single well may enclose adjacent lifts, with the intermediate guide rails supported by steel joists.

The general requirements of well construction given below are specified in BS 2655:Part 6. Several other codes are similar in this respect.

(1) The enclosing walls must be continuous, with openings only for ventilation, landing doors and access panels. The last are necessary for emergency use where the lift is intended to travel non-stop over long distances and should be installed at maximum intervals of 11 m. The strength and rigidity of the enclosure must be sufficient to support the guide rails and landing doors; non-structural panels must have sufficient rigidity to withstand horizontal thrusts without any deformation that might interfere with the movement of the car.

(2) Materials must be incombustible and not degenerate with harmful fumes or gas when fire is applied. The fire resistance of the well structure is usually determined by that required of the building structure as a whole, which is related to the size of the building, its compartmentation and the type of occupancy. In addition, since the well penetrates storeys which, in themselves, may form single fire compartments the walls, doors and access panels can be required to have sufficient resistance to prevent the spread of fire through the shaft. These requirements are specified in the various codes on structural fire protection.

(3) Inner surfaces of the well must be smooth, and any projections or ledges bevelled on the underside. Where car doors open prematurely the facing surface must be vertical and flush.

(4) Access to the shaft, for maintenance, must not be by any opening beneath the counterweight. Where a common shaft encloses two or more lifts, a screen rising at least 2·10 m from the base of the well must separate adjacent lifts.

(5) The well may not contain other building services nor be used as part of the ventilation system. To ventilate the shaft itself, ventilators at the top must have a minimum opening area of 0·1 m² per lift and, with lifts of 2·5 m/s or over, a minimum total opening of at least 3 m².

In the lift motor room conditions must be adequate for engineering maintenance work. Electric lighting, with a minimum horizontal illuminance of 200 lx is required, and there should be a mains power supply other than that to the lift machinery. Surface materials should be dust-free and incombustible; general building regulations may, in addition, require a minimum period of fire resistance to the enclosing walls, floor and roof. Ventilation, and heating if necessary, should restrict the air temperature variation to the range 4–40°C, to ensure stable operation of the machinery.

BS 2655 requires that structural support for the hoisting machinery be calculated on the assumption that the total load is the weight of all fixed apparatus plus twice the weight of those parts of the lift which have vertical motion (including the contract load). Beam deflection should not exceed 1/1500 of the distance spanned across the lift well. The supporting beams must be steel or reinforced concrete.

Safe access to machine rooms must be ensured when this is by trap-doors or from the roof of a building. The door must be kept locked and a warning sign displayed at the entrance.

In conventional buildings the installation of the lift machinery takes place after the construction of the shaft and motor room is complete and the whole made weathertight. Scaffolding is frequently required within the well; there are statutory safety precautions associated with this and with the protection of door openings into the shaft, and these are generally the responsibility of the building contractor. During erection the lift contractor may require the services of other building trades, joiners and bricklayers for instance, the storage of equipment on site, and full electrical power supply from a specified date. Methods of system building and the construction of very tall buildings can require a programming of operations that involves lift installation work starting before the associated building work is complete, and in these cases coordination of the contracting firms is essential from the earliest stages.

3.1.7 Special requirements

(1) *Fire access.* Under certain building codes, lifts are required to be available for use in firefighting. In Britain clauses concerning this are to be found under the requirements of tall building in CP3,[21] Greater London Council regulations[65] and the Scottish Building Standards.[131] A switch (marked Fire Switch and enclosed in a glass-fronted box) should enable firemen to control one lift in isolation from landing call points; the lift should open on to either a staircase enclosure or ventilated lobbies at landings; it may be necessary for the lift to be in a separate shaft from other lifts and to operate on an independent power supply.

(2) *Disabled users.* A lift is the principal means of vertical circulation for many of the disabled. The minimum car size needed to accommodate a wheelchair and an accompanying person has internal dimensions 1100 × 1400 mm, the eight-person 600 kg standard lift of Tables 1 and 3 in BS 2655:Part 3. The standard clear door opening of 800 mm is just adequate with an unobstructed space in the lift lobby of 1500 mm in front of the doors.

Doors that close rapidly or which start to move after remaining open only a brief period can cause distress. It is recommended[63] that lifts providing for the disabled should have a door closing speed not greater than 0·4 m/s in public buildings and 0·2 m/s in residential homes. Preferably doors should have photo-electric control and safety edges; where closing is not delayed by an interrupted light beam the delay before closing should be 7 s.

Lift control buttons and switches should not be higher than 1350 mm for wheelchair users and should preferably be less than 1200 mm. For people with impaired hearing, visual indicators of the imminent arrival of a lift and of its direction of travel should be fixed at landings.

3.2 Paternoster lifts

Paternoster lift cars are suspended at diagonally opposite corners by two endless chains. These are carried on drive sprockets at the top of the shaft and around corresponding idler sprockets in the pit. Four guide shoes on each car engage rails at either side; a horizontal rail at the base of the pit and a 'spearpoint', formed by the two central rails meeting at the top, guide the car during the crossover as the cars reverse. The cars thus travel vertically between the lowest and highest landings, and are displaced sideways at the extremities by a distance equal to the diameter of the sprocket wheels.

The shaft for the adjacent upwards- and downwards-moving cars is, in plan, approximately 2·6 m wide and 1·6 m deep. The floor area of a car, designed to carry two people, is normally just below 1 m^2. The area required for traction equipment above the shaft is approximately double the actual shaft area; the requirements of access space and the accommodation of control equipment lead to a machine room with a minimum size about 4.3 m square. The height of the motor room ceiling above the highest floor served by the paternoster varies with the storey height of the building and is usually between 6·8 m and 7·6 m. The depth of the pit below the lowest landing varies similarly but is usually between 4·6 m and 5·8 m. The pit, to accommodate the guide sprockets and access to them, is required to be both deeper and greater in plan area than the pit of conventional lifts. The overall area needed is close to that of the upper machine room.

The traction motor is mounted above and to the side of the well. Power is transferred to the drive sprockets through a series of reduction gears, a rotating shaft connecting the motor with the sprocket on the other side of the well. A hand winding wheel, for emergency use, is provided. As with other lifts, a brake must be fitted that is automatically mechanically applied when electric power is cut. Figure 3.5 illustrates the principal machine-room equipment before installation.

Car enclosures may be similar in construction to the cars of conventional lifts but, being smaller and travelling at constant speed, the problems of achieving rigidity with minimum weight

Fig. 3.5 Paternoster drive equipment before installation.

are more easily solved. The internal surfaces of walls and the faces of the architrave should be smooth with no projections except hand holds; the internal face of the well opposite to the door opening must be smooth, and the face of the sprocket wheels must not present dangerous surfaces.

Between the floor of a car and the roof of the next car, continuous panels form smooth faces to the landing opening. The panels fold back when pressure is applied and return when released. Similarly, hinged flaps at the sill of each landing opening prevent damage to objects projecting from upward-moving cars. Above the highest landing opening a hinged flap causes the paternoster to stop if displaced by a projecting object. Emergency stop buttons are fitted at every landing and re-starting is effected by a key-operated switch. The car floors and hinged panels should differ in colour from the landing floors and walls.

Under BS 2655 the maximum speed of a paternoster is

limited to 0·4 m/s. It is usual in a new installation for the initial running speed to be 0·3 m/s. The speed is increased as users become familiar with the equipment.

3.3 Escalators and passenger conveyors

An escalator is constructed on a welded steel truss, approximately 1 m deep, that spans diagonally between the two landings. The upper members of the truss lie at the floor levels and along the line of the steps. Within the depth of the frame is the mechanism that carries the moving treads, two endless chains driven around sprocket wheels at the top and bottom of the stair, and the motor, which is usually at the top in the horizontal section of the truss. The moving handrails are driven by separate sheaves within the balustrade panels.

Lifting points should be provided to enable the equipment to be hoisted into position during installation. To make handling easier the structural truss of the escalator may be divided into three sections which are joined on site but most escalators are delivered in one unit. An opening in the building about 3·3 m high is required for the equipment to be brought in. Figure 3.6 shows an escalator before installation in a department store.

The steps are formed of rigid tread and riser units, each having two pairs of rollers. These run on steel angle tracks, attached to the structural frame, the two pairs of tracks varying in their relative positions to form the treads into stepped and flat surfaces. Tread surfaces must be grooved along the line of movement and mesh with comb plates at start and finish of the stairway to reduce the risk of objects becoming trapped. Codes of practice specify in detail the dimensions of grooves and comb.

The handrail, which should move at exactly the same speed as the steps, is subjected to continuous flexure. The usual construction, not dissimilar to a car tyre, is a rubber-covered base of canvas and cording. This curves around a continuous guide fixed to the top of the balustrade and a handrail sheave, above the motor, which gives the traction. Idler sheaves guide the rail at the base of the balustrade and are adjustable to take

Fig. 3.6 An escalator during installation in a department store. The size of the equipment frequently requires that escalators are delivered before the structure of the building is complete
(Reproduced with acknowledgement to the *South Wales Evening Post*)

up slack and allow speed adjustment. A balustrade of toughened or laminated glass can be installed in place of the cheaper solid panels, in which case the rail is driven by rollers within the truss.

The driving unit is usually an a.c. induction motor acting through a worm reduction gear to the drive sprocket. A governor, to disconnect the electricity supply if the speed of the stair becomes significantly greater than the rated speed, may be required and a brake is necessary, mechanically applied and electrically held off. Devices to stop the escalator in the event of a broken step chain, broken drive chain, or reversal of movement are fitted. Stop switches must be installed in the machinery spaces at top and bottom of the escalator, and are necessary on the balustrade for emergency use. All machinery spaces must be secured against unauthorised access and the normal means of starting, stopping and reversing the escalator should be a key-operated switch.

The standard angles of inclination are 30° and 35°. The former is the more common and is almost universally adopted in the USA; a 35° rise is permitted in Britain under BS 2655 provided that the speed is not greater than 0·5 m/s and the rise not greater than 6 m. Some manufacturers also produce escalators with angles of inclination less than 30°. The maximum speed (along the line of inclination) allowed under the British Standard is 0·75 m/s; this is greater than permitted under certain other codes but the speeds found most frequently in practice are 0·45 m/s and 0·60 m/s. A horizontal run of moving treads is necessary at the top and bottom of an escalator to enable passengers to step on and off without difficulty; under BS 2655 this is required to be a distance of 400 mm for speeds up to 0·45 m/s, 535 mm up to 0·65 m/s, and 1220 mm for contract speeds up to 0·75 m/s.

The lower limit of escalator tread width permitted by BS 2655 is 600 mm and the maximum 1050 mm, treads below this range being likely to cause passengers' feet to drag along the balustrade and with greater widths three passengers may stand on each step, the central person unable to reach a handrail. The facing surfaces of the balustrades slope outwards and the widths at hip level, just below the handrail, are greater by

200–250 mm. Standard escalator sizes are based on step widths of 610 mm and 1020 mm (24 inch and 40 inch) giving nominal widths at hip level of 800 mm and 1200 mm. Some manufacturers also produce an escalator of nominal width 1000 mm. The overall width of an escalator is 400–600 mm more than the tread width. BS 2655 requires the handrail height to be between 840 mm and 1040 mm above the line of the step nosings and the rail to extend at least 300 mm beyond the comb-plates at top and bottom.

Passenger conveyors with a moving walkway formed from a series of linked pallets, resembling escalator steps, differ little in construction from escalators, although there is some difference in the application of the traction drive and in the linking of pallets between the products of different manufacturers. In another form of conveyor the walkway is a continuous belt, grooved to mesh with comb-plates, constructed of rubber on a fabric or steel foundation. This may be carried on transverse rollers or a slider bed; it may, alternatively, be stiff in the transverse direction and be supported on edge rollers and guides.

The maximum speed permitted under BS 2655 is 0·90 m/s when the rise does not exceed 8° and 0·75 m/s between 8° and 12°. The handrail should be between 840 mm and 1040 mm above the tread way and extend at least 300 mm beyond the line of the comb-plates. The width limits are the same as those for escalators.

CHAPTER
4
Stairs
and corridors

4.1 Empirical calculations

The dimensions of corridors and stairs are not usually based on calculations of traffic flow. Pedestrian flow rates within buildings are often very small, and the designer determines the width of circulation areas on subjective criteria of lighting and proportion, or the occasional need to move bulky goods. The two principal cases for which calculations can be necessary are:

(1) main pedestrian routes in large public buildings – transport terminals, shopping centres, auditorium buildings;
(2) means of escape from fire.

The cases differ. In the first, the analysis must be based on predicted traffic intensities; it is in these that a considerable error is possible and a high degree of judgment may be required of the architect and the client. In planning escape routes, on the other hand, little is left to the designer. Statutory codes specify densities of occupation to be adopted and rules for designing the escape system, with the result that the provision of escape routes can become a rigid constraint on the design of a large building. The requirements of emergency escape are described in Section 4.2, p. 105.

4.1.1 Corridors

The number of people who can be accommodated within a corridor is a function of the walking speed and the area occupied by each person:

flow rate (P/s) = mean speed (m/s) × mean density (P/m^2)
$$\times \text{ width of route (m)} \quad (4.1)$$

But walking speeds vary with the population and with the physical surroundings. Even within a homogeneous sample – of women shoppers, for example, or of students on a college campus – the range, between the highest speed observed and the lowest, can be equal to the mean value. There tends to be a Normal distribution of walking speeds, the mean varying much more between different samples than the standard deviation.

Many surveys of walking speeds have been reported. The majority have been studies of street design, pedestrian crossings in particular; those of interiors have been mainly surveys within transport buildings or studies of fire escape routes.

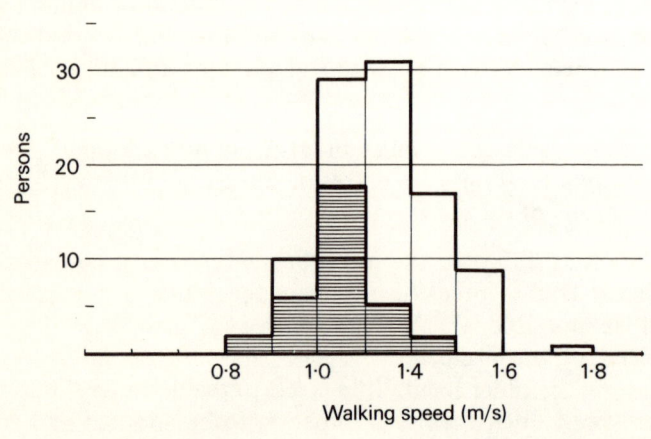

Fig. 4.1 Observed walking speeds in an indoor shopping mall. Shaded area indicates persons walking with one or more others

Differing conditions between surveys introduce hazards into comparing one set of results with another but nevertheless several factors associated with differences in mean walking speed can be identified.

Age and sex differences. Mean walking speeds vary with age. Peschel,[115] in a study of pedestrian crossings on highways, gives a mean speed of 1·1 m/s for children between 6 and 10 years, 1·8 m/s for adolescents, 1·7 m/s for men below 40, and 1·5 m/s for men over 55. Other studies also show a general decline with age in the walking speeds of adults, the decline being more pronounced at ages above 65, but fast and slow walkers are to be found in every age group.

Women tend to walk more slowly than men. The mean values given by Peschel are 1·4 m/s for women below 50 and 1·3 m/s for older women. Fruin,[58] in a survey of pedestrians in bus and rail terminal buildings, found 1·4 m/s to be the average male walking speed and 1·3 m/s the average for women, both values relating to passengers without baggage; Hoel[71] found a difference of about 6 per cent in walking speeds in city centres; in some surveys the differences between walking speeds of men and women has been found insignificant in samples of moderate size.

Groups. The average speed of people moving in groups is generally less than that of people walking alone. In family groups – couples, and parents with children – the whole group tends to walk at a rate close to the normal pace of the slowest member and, among a group of adults, conversation while walking can affect the speed. Peschel found that the mean walking speed of women with young children was 0·7 m/s and, from measurements in an English shopping centre,[147] it was found that the average speed of women with one or more other people was about 20 per cent less than that of women walking alone.

Trip purpose. The enthusiasm of individuals for the journey, the need to hurry or the wish to dawdle, may cause a very large part of the scatter found among any set of measurements of walking speed. It must be taken, in predicting the flow of a crowd, that there will be individual variability which can, in analysis, be treated only as randomness.

But there is some evidence of systematic differences in walking speeds between people engaged in different occupations.

Surti and Burke[139] found that the average walking speed of tourists outside the White House in Washington was 1·0 m/s while that of other pedestrians there was close to 1·6 m/s. Walking speeds of pedestrians crossing highways are greater than in most other circumstances. In the research described by Hoel, walking speeds in Pittsburgh were found to vary with the time of day (being highest between 8 a.m. and 9 a.m.) and with the destination of the pedestrian (people walking to restaurants travelling more quickly than those with business or shopping purposes!). It was also found that walking speeds were higher in cold weather, the data given, between −10°C and 24°C, yielding a regression of approximately −0·5 m/s with each 10°C rise in air temperature. Oeding[110] has recorded that the walking speeds of workers near the gates of a large factory tended to be greater than those measured during the morning peak period in city centres, and that the mean speed of the factory workers was 0·2 m/s greater when leaving work than when arriving. The walking speeds of people attending meetings at sports stadia were found to be slightly less than those in the city centre and the mean speed of shoppers was lower than that of people travelling in the morning peak period.

There is some evidence that, in buildings, hard floor surfaces are associated with higher walking speeds than soft surfaces such as carpet.[122] This is not necessarily related to the physical ease of movement.

Baggage. In uncrowded pedestrian areas there is little difference in the walking speeds of people carrying baggage and those without,[58] but travellers with heavy cases may walk quickly with them for some distance, then stop and rest. With the congestion of heavy flows of people the presence of some with baggage, or trolleys or perambulators, can reduce the overall mean speed.

Gradient. Ramps of 5 per cent or less* have been found to have little effect on walking speeds, either upwards or downwards. Only few data are available on the effect of steeper ramps, but from these it could be inferred that a 10 per cent gradient may reduce the walking pace by 20 per cent and a 15

---

* Gradient per cent $= \dfrac{\text{vertical distance}}{\text{horizontal distance}} \times 100$.

per cent gradient by 40 per cent.[126] The effect of a downward gradient is similar; it has been found that under some circumstances people walk more slowly downhill than uphill, but this has not been observed generally. Less energy is used in moving downwards but the greater control necessary is difficult for the frail and elderly.

Differing flow directions. Heavy pedestrian traffic is hindered when streams of people intersect, and the capacity of a corridor may be less with flow in both directions than with a unidirectional flow. It has been shown[58] that when one stream of pedestrians crosses another the probability of a person having to alter his step is almost 100 per cent when the overall density in the crossing zone is 1·0 P/m^2 but the probability approaches zero when the density is less than 0·25 P/m^2. 'The total capacity at the junction of corridors is smaller than that of the same floor area with unidirectional flow; intersecting streams should be avoided with very heavy traffic.'[94]

The reduction in the capacity of a corridor with two-way flow is also a function of the density. It may be inferred that the effect is not a simple one and that it depends on the walking habits of the particular population, for although in one case a reduction of up to 14 per cent in capacity has been found when one-tenth of the people using a walkway move in the opposite direction to the remainder,[108] in other surveys no significant reduction has been apparent.[94] A tendency for pedestrians to keep to the right of a walkway has been observed[54] and in several surveys it has been noted that people moving in opposite directions may form separate streams near the sides of a corridor. Where it is necessary to maintain discrete channels of pedestrians, the location of entrances and exits to a walkway is found to have a greater effect on behaviour than signs that state, for example, 'Keep to the right'.

Density. The effect of crowding on walking speeds has been measured in several surveys. As the density, the number of people per unit area, is increased the average speed becomes less, walking being reduced to a shuffle at a mean density of 3 P/m^2 and forward movement halted at about 5 P/m^2. The capacity of a corridor, the product of speed and density, is at a maximum just before movement is halted, but these conditions are uncomfortable, movement is slow and any unexpected impedance jams the corridor. The full design values recommended in Table 4.1 are based on the assumption of a density where the flow of people is about 80 per cent of the maximum and the average speed is double that occurring at maximum capacity. Between 0·3 P/m^2 and 2 P/m^2 the reduction of speed with increasing density is, in most cases, linear and the slope such that a doubling of density corresponds approximately with a halving of speed. Empirical curves relating walking speed and density are illustrated in Fig. 4.2.

Walking speeds are dependent, then, on a number of factors

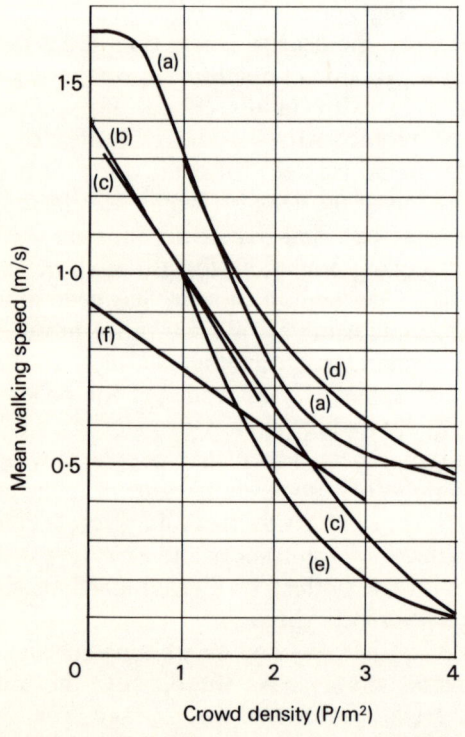

Fig. 4.2 Variation of mean walking speed with crowd density
    (a) Hankin and Wright[68]   (d) Kimura[143]
    (b) O'Flaherty and Parkinson[111]   (e) Togowa[143]
    (c) Older[112]   (f) Foot[54]

related to the physical surroundings and to the subjects themselves. Table 4.1 lists the walking speeds, on level ground, of differing groups of people. These are derived from the surveys referred to but, as the methods of observation and the backgrounds to the various studies have differed considerably, the values must be treated as approximations. Characteristics of the particular population for which a building is to be designed may imply changes to the figures that are related to the several factors discussed earlier.

Table 4.1. Approximate mean walking speeds. Corridor capacity [58, 94, 111, 112, 147]

	Free flow: mean density 0·3 P/m² or less		Full design capacity: 1·4 P/m²	
	Walking speed (m/s)	Limit of corridor capacity with free flow P/min per metre width	Walking speed (m/s)	Corridor capacity P/min per metre width
Commuters, working population	1·5	27	1·0	84
Individual shoppers	1·3	23	0·8	67
Family groups: shoppers with a high proportion of young children or with bulky packages; tourists in circulation areas indoors, or outside near places of interest	1·0	18	0·6	50
Schoolchildren	1·1–1·8 increasing with age	18–32	0·7–1·1	59–92

Under free flow conditions the range of speed in any group may extend, typically, from 0·6 m/s below the mean to 0·6 m/s above. With crowding and all pedestrians moving in one direction the range is very small.

The free-flow conditions correspond with Fruin's 'level of service A' and 'grade 1' of Oeding.[110] Below a concentration of 0·3 P/m^2, interaction between pedestrians is not obviously apparent to those involved and individuals are able to select their own walking speeds, bypassing others who are taking different routes or moving more slowly. A density of 1·4 P/m^2 is the concentration at the 'working maximum flow' given by Hankin and Wright;[68] it corresponds with Fruin's 'level of service E' and Oeding's 'grade 4'. Under these conditions most people walk at less than their natural speed and will be aware of uncomfortable crowding.

For short periods and short distances greater unit flows are tolerable. Small groups causing very tight crowding can be accommodated within a stream of pedestrians provided that at any point these can be seen as only momentary occurrences; they are a normal feature of heavy traffic. Similarly a crowd can pass through a short section of narrow corridor, up to about 3 m long, at a higher density than is feasible over a greater distance, provided that significantly wider spaces occur before and after the restriction.[94]

If a staircase or corridor is less than 1·2 m wide, the maximum capacity is not proportional to the width: eq. (4.1) ceases to apply, because the shoulder widths of individuals affect the rate of movement where the way is too narrow for two people to walk abreast with ease. A single lane width is, in many cases, taken as being 21–22 inches (about 0·55 m) and this is a useful guide when people are moving slowly under controlled conditions, but greater lane widths have been recorded in other circumstances. Narrow streams of traffic can occur when a corridor is divided by handrails. This usually precedes the formation of queues – service at a counter or controlled entry to a building – and the flow of people is not continuous. Where there is an unobstructed stream it may be inferred from the few available measurements that the capacity of a single lane of 0·55 m is about 0·5 P/s, a headway of two seconds between people, but this can be expected to vary significantly with the population and with the nature of the route. Small increases of width do not affect the capacity, since faster-moving pedestrians remain unable to pass the slower, but

Table 4.2. Approximate reductions from the effective width of a walkway (m)

Single queue along side of corridor	1·2*
Persons seated on bench along wall	1·0
Coin operated machines	depth of machine plus 0·6 for first person and 0·4 for each additional user
Waiting pedestrian with baggage	0·6*
Shop windows	0·5–0·8, varying with the interest shown in the goods, and with their size
Small fire-fighting appliances	0·2–0·4
Wall-mounted radiator	0·2
Rough or dirty building surface	0·2

* See section 4.1.2.

as the width approaches 1·2 m overlapping lanes form and the rate of flow is increased.

Space at the sides of corridors may be wasted from the circulation area. Firefighting appliances and other projecting objects can reduce the effective width of a corridor, and the capacity of the route may be reduced by the presence of people who are not part of the principal flow. Some typical dimensions are listed in Table 4.2.

Where only sparse flows of people are anticipated (as in most cases) the width of a corridor may be determined by the

Table 4.3. Minimum width of straight corridors (m)

Two men passing abreast	1·2
Porter with baggage trolley	1·0
Man carrying baggage	1·0
Woman with perambulator	0·8 (1·2 with young child alongside)
Man with crutches	0·9
Wheel chair	0·8*

* See section 4.3.

Wheeled vehicles required additional space when turning, the greater the length of the vehicle the wider the lane necessary. At a right-angled junction, wheelchairs and most perambulators can be turned if either the entry or the exit corridor is twice minimum width for forward movement.

width of wheeled vehicles and the ease by which persons can pass each other. Table 4.3 gives a number of examples.

4.1.2 Doorways

The flow capacity of doorways has been measured by several investigators concerned with escape routes from fire. Very high rates of flow, up to 5 P/s per metre width have been recorded when subjects are hurried and pushed through an exit doorway, but with normal pedestrian movement the rate is about 1 P/s.

Table 4.4. Approximate capacity of openings [25, 58, 102, 143]

	(P/min)
Gateways and other clear openings	60–110 per metre width
Single swing door (0·9 m)	40–60, increased by 50% if fastened open
Revolving door	25–35 in one direction. This is doubled if leaves collapse to give two openings
Waist-high turnstile:	
with free admission	40–60
with cashier	12–18
operated with single coin	25–50

Several codes governing means of escape specify the capacity of doorways in terms of unit widths, 0·55 m or so, but except when people are moving in disciplined files there is little evidence that the capacity of an opening is not directly proportional to the width.

4.1.3 Stairways

A crowd of people moves upstairs more slowly than along a level corridor. The mean speed of the London Underground passengers observed by Hankin and Wright[68] was 0·8 m/s (along the line of slope) under free-flow conditions; this was about half the level walking speed measured. An increase in density, however, has less effect than in a corridor. At a density in plan of 2·0 P/m² the upward speed of 0·6 m/s was three-quarters that of level movement.

Speeds downstairs are normally greater than upstairs, sometimes 10 per cent more, but at high densities, with hurrying and pushing, the reverse can be true. The walking speed of men is greater in both directions than that of women, and people in groups are slower than individuals. As with level walking speeds, the variation within each category is large and the scatter of survey results great enough to mask differences between separate categories of people unless large samples are taken.

Recommendations for stairway design capacity vary widely in the literature. For fire-escape stairs in buildings a value around 1·3 P/s per metre width is adopted in several codes, but this is much greater than found in normal use and for stairways in city subways, for instance, rates less than half this figure have been proposed.[110] It is clear that when the occupants of a building are walking down a stair under controlled conditions its effective capacity can be higher than 1·3 P/s per metre width, but in ordinary conditions such crowding is considered unpleasant. Two-way traffic, unless channelled along separate sides of a stair, causes lower speeds than one-way flow, and the presence of children or elderly people can retard the flow in crowded conditions.

The values given in Table 4.5 under 'full design capacity' describe conditions that occur in railway stations or pedestrian subways during peak-hour crowding. Virtually every person using the stairs is hampered in his stride and movement can be halted by minor obstructions. Two-way flow can occur only with great difficulty. The figures under 'free-flow conditions' give the limits of traffic flow where individuals are able to maintain their own paces with ample room for passing slower walkers and avoiding those travelling in the opposite direction.

It is important that the combination of the riser and tread dimensions of a stair should lie within the natural gait of the users and this, taking the average pace of an adult climbing stairs to be 0·6 m, is the origin of the traditional requirement that the sum of the tread plus twice the riser height should equal that value. There is no basis for demanding exactness here, and Fitch[51] recommends combinations of riser and tread dimensions from 0·10 × 0·36 m to 0·18 × 0·28 m, conclusions reached after experimentally examining gait, freedom from mis-steps and energy expenditure. For domestic staircases within a standard storey height of 2·6 m, it was found by

Table 4.5. Approximate mean speeds of movement up stairways. Stair capacity [25, 58, 94, 102, 110]

	Free flow: mean plan density 0·6 P/m² or less		Full design capacity: plan density 2·0 P/m²	
	Speed along slope, (m/s)	Limit of stair capacity with free flow (P/min per metre width)	Speed along slope (m/s)	Stair capacity (P/min per metre width)
Young and middle-aged men	0·9	27	0·6	60
Young and middle-aged women	0·7	21	0·6	60
Elderly people, family groups	0·5	15	0·4	40

Ward[5] that a stair of fourteen risers with a going 265–270 mm was 'ideal' but within a limited plan area a staircase with 12 risers and a going 241–245 mm was satisfactory.

The safety of stairs has been studied during analysis of accidents and by laboratory experiments. More accidents occur in descending than in climbing stairs, and Fitch has shown that the number of mis-steps increases as the depth of the tread is reduced. Ward illustrates how, with smaller treads, subjects place a lesser proportion of the shod foot on the stairs, the ankle being twisted to place the foot diagonally across the steps when descending.

The rate of accidents is high when steps in public places are difficult to see. Quite apart from the level of illumination, this is likely to occur when one or two isolated steps are approached from above and there is no change in the floor surface finish. It is frequently specified in codes of practice that no flight should have less than three risers. Short flights in an otherwise unbroken floor plane should be marked with variation in the colour or material of the floor finish or with prominent handrail.

Although it has been found that people when walking on the level tend to choose a pace that gives the minimum expenditure of energy,[36] this does not necessarily apply to stair climbing. A short flight of steps is seldom climbed at a speed that gives the lowest physiological cost.[51] The actual energy necessary to ascend a given vertical distance varies with the stair proportions. The total amoung of energy used can become smaller as the stairs increase in steepness, but the rate of energy expenditure rises. The rate increases, also, as the sum of the tread and riser becomes larger, and a low riser and large tread can be more tiring than a stair of moderate proportions.

The energy required per metre rise when walking up a ramp decreases as the angle increases up to 17° but becomes greater at higher angles; to ascend a given height more energy is needed for a ramp than for a flight of stairs of normal proportions. Considering both the horizontal and vertical distances, however, the use of a ramp is physiologically more efficient at low angles than climbing a stair and walking a comparable distance on the level.

4.1.4 *Waiting areas*

Individuals waiting in railway stations and in similar public buildings tend to choose locations where there is physical protection from the main flows of pedestrians but where they can see, and be seen by, people entering the space.[136] Where there is a number of people waiting for service of some kind, the form of the group can vary. It is convenient to distinguish between two categories: bulk queues and linear ordered queues.[58]

The first of these occurs on a railway platform and in a theatre bar during the interval; there is little relationship between the order of arrival and the location of individuals in the crowd. At 2·0 P/m^2 a crowd seems dense to those within it but where there are physical constraints, as in a lift car, people will stand more closely together. The mean density of standing passengers in modern London Transport underground rolling stock was found to reach 7·2 P/m^2 in the doorway area[149] although such crowding is not often tolerable. The maximum density at which people are comfortable while standing waiting is in the range 1·0–1·5 P/m^2,[58] but this varies with the population and the circumstances. Passengers standing with bulky luggage in London railway stations have been found to cluster at about 1·0 P/m^2.[147] A density not greater than this should be adopted when designing bulk queueing areas within which some circulation will occur, and an overall density of about 0·4 P/m^2 in public waiting areas has been suggested.[5]

An ordered queue before a service point, such as a ticket window, may not take the same form in plan on different occasions. There is a tendency (observed in Britain, at least) to queue alongside a wall, but if the queue grows backwards to an obstruction or if a few people come to stand directly in front of the service area the queue can become a broader line extending outwards. Stilitz[136] described how a queue of people at a London Underground ticket office tended to form along the line of approach to the window and how, when a queue interfered with a cross flow of people, the flow gave way if it was sparse but the queue itself was deflected by a heavy flow.

The width of a waiting line can be limited with barriers to 0·6 m but a queue that is not constrained will spread. The mean width of a bus queue of six people has been recorded as 1·2 m;[111] the width of a long queue at the ticket window of a London Underground station during a holiday weekend was found to vary between 0·6 and 1·6 m with, on average, 8·0 passengers in a 3 m length of the queue; a long static queue in a London main-line terminus, passengers standing in groups with heavy baggage, varied in width from 1·5 to 3·0 m, the mean number of passengers in each 3 m section being 6·3.[147]

Waiting areas are an essential component of any complex circulation system. A stream of traffic with a mean flow rate that is constant over a long period normally exhibits transient flow rates of high intensity and also periods with long gaps between successive arrivals. When the characteristics of service and arrivals can be predicted, so can the probability distribution of the number of people queueing at a service point; there is a considerable body of literature on the subject, although the number of publications on mathematical queueing theory is much greater than the number describing its application. In essence, if there is any randomness, in arrivals or in service, the capacity of each of the various elements of a circulation system must be greater than that needed to accommodate a constant flow and, if the arrival and service mechanisms are independent of each other, no matter how great the difference between the mean rate of one and the mean capacity of the other, there remains a small but finite probability of queueing occurring.

The movement of people in a building is not, though, always independent of the service pattern. For this reason the temporary restriction of a flow of people may not cause an observable queue. The service ahead can be anticipated by hurrying to fill a gap or by slowing the walking pace when congestion is apparent. This can improve the rate of service and is subjectively desirable; there is an argument, then, for making areas where congestion can occur visible for some distance along the preceding route.

4.2 Escape from fire

Means of escape must be foolproof. Any person should be able to leave a large and complex building rapidly and without difficulty, even though the electric lighting may fail and a crowd of people may panic. There are some features that are usually essential:

(1) The routes must be apparent throughout the building. In some cases emergency lighting and illuminated signs are neces-

sary but, in general, the need to read directions should be avoided: an escape route should lie in the direction a person would follow naturally in an emergency.

(2) Two or more routes are required where fire could make access to a single exit dangerous.

(3) The period during which occupants are exposed to fire should be short. Where moving out of the building may take several minutes, the corridors and stairs used for escape must be protected from smoke, fumes and excessive heat, and the distance limited from any point in the building to a protected area.

(4) A crowd moving along an escape route must not be impeded. Doors should swing open in the direction of flow, doorways should not be narrower than the corridors that lead to them, and corridor widths should not be reduced by stored goods or other objects. At the ground floor the route must lead unmistakably to the final exit doorway.

It is assumed in Britain (and, in most cases, elsewhere) that mechanical installations cannot be relied upon to provide escape routes within buildings. The capacity of lifts and escalators is usually insufficient, and the probability of their failing when a fire occurs is not insignificant. It may be necessary for firemen to use one or more lifts for access and, in very large buildings, for the normal occupants to use lifts that serve storeys remote from the fire but, to evacuate a dangerous area quickly, a system of protected stairs and corridors is required.

There is no general solution to the problem of providing means of escape for the disabled. Almost any building may be used by the physically handicapped, the very old or the very young, and this must not be overlooked when making assumptions on the mobility of the occupants. It may be reasonable to assume that fellow workers in a multi-storey office building could assist an individual who is normally unable to use stairs, but such cannot be assumed when designing places of public assembly. It is necessary, when designing, to consider the locations to which these special groups of people will require access, and to examine particular attributes of the routes to places of safety. Although the concept is embodied in almost all existing regulations, relying upon non-mechanical means of escape is not ideal, and difficulties are found particularly when considering the disabled.[14]

Safety from fire in buildings is the subject of extensive statutory control. This lies in general documents covering building standards, as for instance the London Building Acts, the Building Standards (Scotland), the Building Regulations, and the Fire Precautions Act, 1971. It is dependent also on statutes covering specific uses of buildings, such as the Factories Act, 1961, the Offices, Shops and Railway Premises Act, 1963, and various acts relating to entertainment and the use of hazardous materials. In Britain, also, the local Fire Authorities have considerable scope in interpreting the intentions of some of these measures and their application can vary from district to district. To the architect there is no substitute for knowledge of the regulations themselves and of local constraints on building.

For an example of a fire-escape system and the particular requirements of a number of codes, the routes within an office building are illustrated in Fig. 4.3. It will be seen that there are two enclosed stairways sufficiently far apart to provide effective different escape paths on upper floors but which, under the relevant British Standard Code of Practice (CP3:Chapter IV:Part 3:1968)[21] must not be separated by more than 61 m. Where a room opens on to a main corridor escape is possible in either direction. For open plan areas there is a rule, given for guidance under the Offices, Shops and Railway Premises Act,[43] that from any point on the floor straight lines drawn to the two stairs should form an angle at least 45°. The same rule is given in CP3 for the spacing of doors to individual offices when the distance between any part of the office and the door exceeds 12 m. It should not be necessary for the occupants of an office to pass through more than one other office to reach safety, and rooms that provide access to an escape route should be in the same occupancy.

The maximum permissible distance from any point on an upper storey to any exit door from the storey is 46 m (CP3), which corresponds with an escape time of $2\frac{1}{2}$ min at a mean walking speed of 0·3 m/s or a delay of about $1\frac{1}{2}$ min from the

108 The Design of Interior Circulation

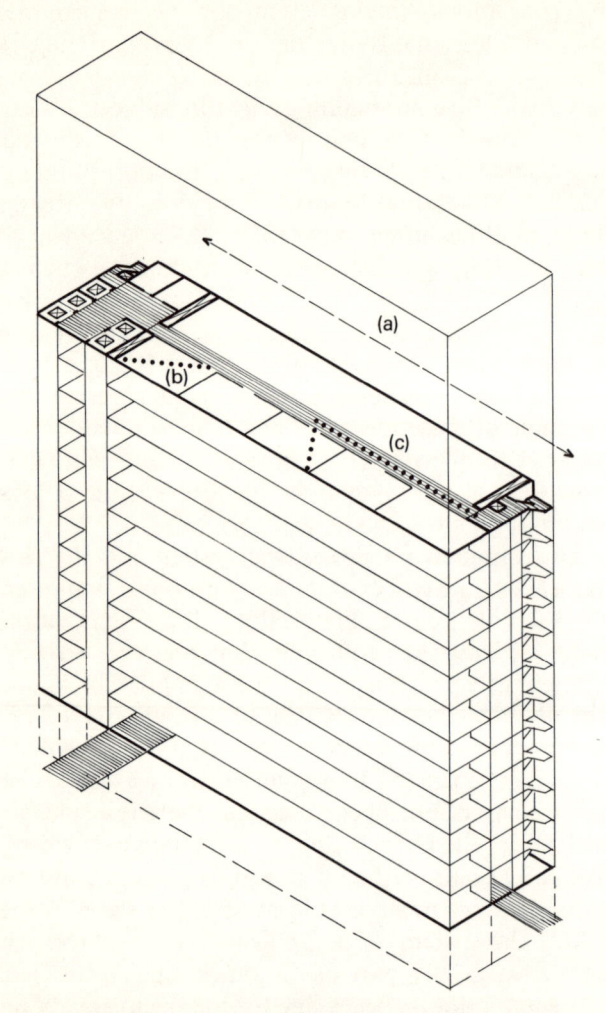

Fig. 4.3 Principal limits on travel distance in tall office buildings
(CP3: Chapter IV: Part 3: 1968)
(a) Maximum separation of escape stairs, 61 m
(b) Maximum distance to a single exit, 12 m
(c) Maximum travel distance from any point in an office to an exit door from the storey, 46 m

sounding of an alarm and a walking speed of 0·8 m/s. A maximum of 31·25 m is required under the Scottish Building Standards; under the Greater London Council code[65] a direct distance not greater than 30 m is permitted and a maximum travel distance one and a half times this. From the American code for safety to life from fire[107] an unprotected distance of 61 m is permitted in a building with sprinklers and 92 m where these are installed. In any part of an office building above first-floor level, where escape is possible in only one direction, a maximum travel distance of 12 m is permitted by CP3 and the Greater London Council code, 12·5 m by the Scottish Building Standards, and 15 m by the American life safety code. In general the travel distances permitted in office buildings are greater than those allowed in institutional buildings, places of assembly and shops.

To design the later stage of the escape routes an estimate is required of the number of people that are to be accommodated in the building. The occupancy of commercial buildings is often difficult to predict — speculative office projects and large shops in particular — and 'occupant load factors' are listed in codes of practice. These are average values of the area occupied per person; the number of people to be assumed when the exact occupancy is unknown is found by dividing the internal floor area by the load factor, some examples of which are given in Table 4.6. The figures are useful in estimating day-by-day traffic in buildings as well as the requirements of fire escape although, by the nature of their original purpose, they lead to an overestimate of the population rather than an underestimate.

Widths of stairs and exit doors are found by assuming that one escape route may be blocked and the number of occupants of a storey is divided between the remaining routes. Most codes contain tables of minimum staircase width against the number of people per floor; in CP3 it will be found that if the building illustrated in Fig. 4.3 has sixty persons on every floor the stairs and the doors leading to them must be at least 1·07 m wide. The Scottish Building Standards require a calculation of the width necessary to clear the floors in $2\frac{1}{2}$ min with a discharge rate of 40 P/min per 530 mm width of stair; at all upper floors

Table 4.6. Occupant load factors (m²/P) to be used when exact occupancy is not known

Life Safety Code[49] (original in ft²/P)		Scottish Building Standards[131]		Other British Codes	
Places of Assembly	1.4	Assembly halls, movable or no seating	0.5	Closely seated audience[c]	0.5
Areas of concentrated use without fixed seating	0.7	Bars	0.5	Dance halls[c]	0.55
Standing space	0.3	Grandstands without fixed seating	0.5	Exhibitions[c]	1.5
		Clubs	0.5		
		Concourses, crush halls and queueing lobbies	0.7		
		Dance halls	0.7		
Stores, street floor and sales				Shops and showrooms[c]	7.0*
basement	2.8*	Shops trading in common consumer goods, basement ground and upper storeys	1.4*	Supermarket and bazaar type stores[c]	2.0*
Other floors	5.6*	Shops specialising in more expensive trades	1.9*	Departmental stores[c] main sales area	2.0*
Storage, shipping	9.3*	Shops for personal service, including hairdressing	7.0*	sparsely occupied areas	7.0*
			1.9	Shops trading in common consumer goods[a]	1.9*
				Specialised shops[a]	7.0*

Educational occupancies:			
Classroom area	1.9	Reading rooms, writing rooms	1.9
Shops and other vocational areas	4.6	Libraries, museums, art galleries	4.6
		Staff rooms	1.1
		Teaching rooms, escape routes to be based on maximum number of children for which each room is designed[b]	
		Dining rooms and gymnasia[b]	0.9
		Assembly halls[b]	0.45
Office, factory and workroom	9.3*	Offices, storeys not divided into rooms	5.1
		Factory shop floors – workrooms and storage	4.6
		Offices[a,c]	9.3* 10.0*
		Factories[c]	10.0*
Hotel and apartment	18.6*	Dormitories and bedrooms other than in private dwellings	4.6
		Dining rooms, cafes, common rooms	1.1
		Kitchens	9.3
		Lounges	1.9
		Restaurants and lounges[c]	1.0–1.5*
		Bars[c]	0.3–0.5*
		Bedroom floors in hotels: occupancy based on number of bedrooms[c]	

* Gross internal area. All other figures apply to net area.
[a] British Standard Code of Practice. CP3: Chapter IV: Parts 1, 2 & 3.[21]
[b] Department of Education and Science. Building Bulletin 7.[42]
[c] Greater London Council Code of Practice: means of escape from fire.[65]

except the highest the stair must be sufficiently wide to carry the population of two floors less the number that can be accommodated standing on one storey of the staircase. The stair widths required in the American life safety code are, in general, based on a flow rate of 45 P/min per unit width of 560 mm.

The detailed design of staircases is specified in each code. Maximum and minimum dimensions for riser and tread are given and, in many cases, so is a formula for relating the tread size to riser height. Most codes describe the circumstances in which winders, stairs with open treads and other unconventional stairs are permitted, and these vary. Overall dimensions of a staircase are normally specified with a minimum and maximum number of successive steps and the required headroom. In the office building illustrated it would be required, under CP3, that risers be not more than 190 mm, treads not less than 254 mm measured from riser to riser, and that each flight has not less than three and not more than sixteen risers. The headroom should be 1980 mm minimum above the line of nosings. The construction of stairwells, in terms of fire resistance and the means of ventilation, is covered in every code.

Escape stairs must lead directly to safety outside the building, clear of any smoke or fire that could emanate from a basement and clear of doorways to any high fire risk areas. If a corridor links the stair with the final exit doorway this should have the same resistance to fire as the stairway enclosure, and any hallway that serves as an escape route should not contain a potential fire hazard. Two or more escape stairs must not terminate in the same ground-floor enclosure, and it is essential that stairs from upper floors do not continue directly into a basement.

The floor plan of the building illustrated is determined, in part, by the provisions for firefighting which are required by CP3 for office buildings with storeys above 18·3 m:

(1) One or more passenger lifts must be available for the use of firemen. In the ground floor lobby a switch is required that will enable firemen to control the lift without interference from the landing call buttons. This is normally enclosed in a glass-fronted box to prevent unnecessary use.

(2) At least one of the fire-escape stairs must be available for use as a firefighting access stair. A lobby of minimum floor area 5·5 m² must separate the stair from the main floor area at each storey. The lobby must have the same degree of structural fire protection as the staircase and it must be ventilated with openable windows and a small permanent opening. Firefighting hydrants and rising mains are normally placed within the lobby. The stair itself must have windows at each storey and permanent ventilation at the top of the staircase shaft. Stairs serving basements must be linked with smoke outlet ducts that are independent of other ventilation systems.

(3) At ground level there should be direct access from the street to both the firefighting access stair and the firemen's lift. The staircase must be continuous throughout the upper storeys.

4.3 Special requirements of the disabled

Designing a building for the safety and comfort of the handicapped requires attention to small details of the interior. The accessibility of the building as a whole, though, can be affected by decisions made during the early stages of planning, particularly on the treatment of the site around the building and on the systems of vertical circulation upon which the internal design is based.

Handicapped people do not form a single group whose requirements are uniform. Their needs vary, and features of buildings that are helpful to those with one type of disability can cause difficulty for others. In the design of a public building the following conditions may need to be considered:

(1) *Sensory disablement.* The blind and the partially sighted, and those with impaired hearing.

(2) *Locomotory and manipulatory disabilities.* People who are confined to wheelchairs, and those who are able to walk, but with difficulty. The latter category in most cases contains the greater number of people and, as well as those with limb injuries and the partially paralysed, includes those with conditions associated with ageing, notably heart and respiratory conditions and arthritis.

(3) *Mental handicaps*. A diverse group including those with hereditary disorders such as epilepsy, patients with intellectual and emotional disorders, and in which may be included those affected by alcohol and other drugs.

The categories are far from mutually exclusive and it is to be expected that many of the aged and of those with congenital disorders will carry more than one type of handicap.

In Britain, under the Chronically Sick and Disabled Persons Act, 1970, buildings to which the public are admitted are required to have parking, means of access and sanitary conveniences available for the disabled, so far as it is both practical and reasonable in the circumstances; the same is required of school and university buildings. A sign indicating that provision is made for the disabled must be displayed outside buildings where these conditions apply.

4.3.1 *Corridors and waiting areas*

Disabled people frequently dislike crowded spaces and are afraid of being jostled. It is likely that people with impaired mobility are sometimes deterred from using public areas if they appear densely occupied. The anticipated use of wheelchairs can determine minimum corridor widths, and the British Standard Code of Practice, CP96 : Part 1 : 1967,[22] requires that corridors to give access to parts of buildings used by the disabled are at least 1220 mm wide. For wheelchairs to pass each other at least 1650 mm is preferable, and this width should be provided where it may be necessary for a chair to be turned through 180° in the corridor.

Railings or other projections that commence above hip level may be undetected by a blind person with a cane and must not occur. Doors that open outwards into circulation areas can also be dangerous. Where this is unavoidable, as in fire escape routes, either the doors must be recessed in a bay or there must be means of preventing people from walking nearer than one metre to the wall in which the doors are set.

Floor surfaces should be non-slip and smooth; the surface of outdoor pavements should not be loose, and external ramps and steps should be protected, where possible, from ice and snow. The use of soft sound-absorbing surfaces on walls and ceiling, or of a floor carpet, may make the mobility of the blind more difficult. In buildings used extensively by those with impaired sight, maintenance of corridors may be aided by providing a deep skirting, to reduce wear by canes, and a wall surface below 1·2 m that is simple to renew.

Seating for the ambulant disabled should be available in waiting areas. For those with stiff hips a narrow shelf type of seat 660 mm high is preferred for short periods; deeper and lower seats can be difficult to rise from.

4.3.2 *Stairs*

Flights of steps are impassable to unaided wheelchair users but some ambulant disabled prefer steps with low risers to ramps. For external steps CP96 requires that risers do not exceed 165 mm and preferably be 150 mm high; the going should be at least 280 mm. Goldsmith[63] recommends a maximum riser height of 145 mm with a minimum going of 280 mm and preferably 370 mm or more. The limiting dimensions of internal stairs given in CP96 are 165 mm maximum riser with 240 mm minimum going, a gradient of approximately 35°.

Open risers, winders and variations in step dimensions must be avoided. Nosings of internal stairs should lap between 20 and 25 mm over the back of the step below, and abrupt nosings are unsatisfactory. No single flight of stairs indoors should rise more than 1830 mm[22] and outdoors, where there may be no protection from the weather, the rise should not exceed 1220 mm. Stairs interrupted by frequent landings can be easier for the elderly than long flights, but single steps and flights of only two steps can be dangerous, as can small steps less than 80 mm high.

Stairways should not be planned so that a door opens directly on to the top of the flight and, with a corner in a corridor at the top, the highest riser must be at least 305 mm from the point in plan where the wall returns. To ensure that the blind have warning of approach to a stair, a light self-closing door to a corridor at the head of a staircase can be of value.

To be distinguished by the partially sighted, changes of floor plane – at steps or at the start of ramps – should be marked by a change in the floor surface pattern or colour, and it is helpful if stair risers differ from the treads. For the blind it is an aid if the surface material of the steps extends about 1 m from the head and foot of a staircase.

4.3.3 *Ramps*

Excepting very short ramps the greatest gradient should be 1:12 ($4\frac{1}{2}°$) and a ramp should not be more than 9 m long unless the gradient is less than 1:20 (3°). Steeper ramps are especially dangerous for the semi-ambulant. There should be a level platform at least 1·2 m long at the top of a ramp, particularly with a high gradient, and adequate manoeuvring space with good visibility must be provided at the base. Level rest-platforms may be desirable in very long ramps but these are best provided at changes of direction. In general, variations of gradient can cause difficulty. Ramps should not be narrower than 1·2 m.

4.3.4 *Handrails and barriers*

Rails should be fitted on both sides of stairways and on both sides of ramps steeper than 1:20. They should continue around landings and extend at least 305 mm beyond the lines of the highest and lowest nosings. The height should be 840 mm above stair nosings or 910 mm above the surface of a ramp, but these dimensions are a compromise between the needs of different individuals and between the differing requirements of upward and downward movement. In some cases, for children or for the elderly, an additional lower handrail at 700 mm can be useful. A width between handrails on stairs of 900 mm allows elderly people to hold both sides. Rails should not be rectilinear or complicated in cross-section; suitable forms are illustrated by Goldsmith and in CP96.

Where a drop occurs, at the side of a ramp or elsewhere, a low kerb (51 mm or more) should be fitted, together with a balustrade at least 910 mm high.

4.3.5 Doors

The clear width of door openings must be at least 785 mm. Wheelchair users need to position the chair to move at right angles through this minimum width; a space at least 965 mm from the face of the door is required to do this if the door opens away from the user and 1690 mm if the door opens inwards. A lesser depth can be allowed with a greater opening width. In institutional buildings the use of large, non-selfpropelling wheelchairs may require greater dimensions. Wheelchair users also need space beside the door handle and the door on this side must be at least 380 mm from a wall that returns back at a right angle from the plane of the door. The maximum height of a door handle should be 1060 mm. Door knobs are more difficult to use, for the elderly, than handles.

Doors with a two-way swing can be hazardous in swinging back violently and, if a 180° opening is essential, a centre check should be fitted to prevent this. Revolving doors must be supplemented with an adjacent side-hung or sliding door. Raised thresholds are unsatisfactory but, if unavoidable, should not exceed 20 mm in height. Frameless glass doors are a hazard to the partially sighted; the plane of the glass must be distinguished by an applied marking between waist level and eye level. Similarly, large windows with sills near floor level can be mistaken for openings and should be fitted with a prominent rail or other barrier.

Bibliography

This includes books and papers on circulation routes in buildings, lift selection and calculation, and pedestrian movement. There are selected texts on the mechanical equipment of lifts. Other entries apply to the numbered references in the text.

1. Adams, G. R., *The design of buildings for the blind*, Unpublished report, 1967 (Library of the Royal Institute of British Architects).
2. Adler, R. R., *Vertical transportation for buildings*, Elsevier, New York, 1970.
3. Agraa, O. M. and Whitehead, B., A study of movement in a school building, *Building Science*, 2, 279–89, 1968.
4. Agraa, O. M. and Whitehead, B., Nuisance restrictions in the planning of single-storey layouts, *Building Science*, 2, 291–302, 1968.
5. A.J. Handbook: building services and circulation, *Architects' Journal*, **151**, 1970. (Published in instalments.)
6. American Society of Mechanical Engineers, *American national standard safety code for elevators, dumbwaiters, escalators and moving walks*, ANSI A17.1–1971, New York, 1971.

7. American Standards Association, *American standard specifications for making buildings and facilities accessible to, and usable by, the physically handicapped*, New York, 1961.
8. Annett, F. A., *Electric elevators*, 2nd edn., McGraw-Hill, New York, 1960.
9. Armour, G. C. and Buffa, E. S., An heuristic algorithm and simulation approach to the relative location of facilities, *Management Science*, **9**, 294–309, 1963.
10. Atkinson, G. A. and Phillips, R. J., Hospital design: factors influencing the choice of shape. *Architects' Journal*, **139**, 851–5, 1964.
11. Bailey, N. T. J., On queueing processes with bulk service, *Journal of the Royal Statistical Society*, B**16**, 80–7, 1954.
12. Bailey, N. T. J., Operational research in hospital planning and design, *Operational Research Quarterly*, **8**, 149–56, 1957.
13. Barney, G. C. and dos Santos, S. M., *The design evaluation and control of lift systems*, Lift design partnership, Bolton, 1974.
14. Bazjanac, V., Another way out? *Progressive Architecture*, (4) 88–9, 1974.
15. Bedford, R. J., Lift-traffic recording and analysis, *G.E.C. Journal*, **33**, part 2, 69–77, 1966.
16. Bird, E. L. and Docking, S. J., *Fire in buildings* (Chapter 5, means of escape), Adam and Charles Black, London, 71–88, 1949.
17. Black, F. W., Interdepartmental traffic in non-teaching acute general hospitals, *Architects' Journal*, **143**, 701–7, 889–92, 1966.
18. Braf, P-G., *The physical environment and the visually impaired*, ICTA Information Centre, Fack, S-16103, Bromma 3, Sweden, 1974.
19. Brend, H. J., *Means of escape in case of fire*, Pitman, London, 1952.
20. British Standards Institution:
    BS 2655:   *Lifts, escalators, passenger conveyors and paternosters.*

Part 1 : 1970. *General requirements for electric, hydraulic and hand-powered lifts.*
Part 2 : 1959. *Single-speed polyphase induction motors for driving lifts.*
Part 3 : 1971. *Arrangements of standard electric lifts.*
Part 4 : 1969. *General requirements for escalators and passenger conveyors.*
Part 5 : 1970. *General requirements for paternosters.*
Part 6 : 1970. *Building construction requirements.*
Part 7 : 1970. *Testing and inspection.*
Part 8 : 1971. *Modernization or reconstruction of lifts, escalators and paternosters.*
Part 9 : 1970. *Definitions.*
Part 10 : 1972. *General requirements for guarding.*

21. British Standards Institution:
    CP3 : Chapter IV : *Precautions against fire.*
    Part 1 : 1971. *Flats and maisonettes (in blocks over two storeys).*
    Part 2 : 1968. *Shops and department stores.*
    Part 3 : 1968. *Office buildings.*

22. British Standards Institution:
    CP96 : Part 1 : 1967. *Access for the disabled to buildings.*

23. British Standards Institution:
    CP407 : 1972. *Electric, hydraulic and hand-powered lifts.*

24. Browne, J. J. and Kelly, J. J., Simulation of elevator system for world's tallest buildings. *Transportation Science*, **2**, part 1, 35–56, 1968.

25. Bruce, J. A., The pedestrian, in Baerwald, J. E. (ed.), *Traffic engineering handbook*, 3rd edn., Institute of Traffic Engineers, Washington, pp. 108–41, 1965.

26. Building Research Station, *Lifts in large buildings* (Proceedings of a seminar held on 15th June 1966), Garston.

27. Burberry, P., Towards a common theory of movement, *Building*, **216**, 3/77–100, 8/103–6, 1969.

28. Carstens, R. L. and Ring, S. L., Pedestrian capacities of shelter entrances, *Traffic Engineering*, **41**, 38–43, 1970.

29. Cassidy, B., Stairways, ladders and ramps, *R.A.I.A. Handbook*, section (24), instalment 1. Royal Australian Institute of Architects, 1968.
30. Cavagna, G. A. and Margaria, R., Mechanics of walking, *Journal of Applied Physiology*, **21**, 271–8, 1966.
31. Clarke, D., Lifts in tall buildings, *Architects' Journal*, **156**, 327–30, 1972.
32. Closs, G. D., Lift control – conservatism or progress? *Electronics and Power*, **18**, 308–10, 1972.
33. Computers in building: Planning accommodation for hospitals and the transportation problem technique. *Architects' Journal*, **138**, 139–42, 1963.
34. 'Contactor', *Electric lifts* (Electrical Engineer Series, vol. 14), Newnes, London, 1941.
35. Corlett, E. N. et al., The design of direction finding systems in buildings, *Applied Ergonomics*, **3**, part 2, 66–9, 1972.
36. Corlett, E. N. et al., Ramps or stairs: the choice using physiological and biomechanic criteria, *Applied Ergonomics*, **3**, part 4, 195–201, 1972.
37. Costonis, J. J., *Space adrift: landmark preservation and the market place*, University of Illinois Press, Urbana, 1974.
38. Council of Europe, Committee of Ministers, Resolution AP(72)5: On the planning and equipment of buildings with a view to making them more accessible to the physically handicapped, 1972.
39. Courtney, R. G. and Davidson, P. J., *A survey of passenger traffic in two office buildings*, Building Research Establishment Current Paper 67/74, 1974.
40. Craig, C. N., Factors affecting economy in multistorey flat design, *RIBA Journal*, **63**, 240–9, 1956.
41. Davidson, P. J., *A survey of passenger traffic in a Croydon office block*, Note no. N85/75, Department of the Environment/Building Research Establishment, 1974.
42. Department of Education and Science, Building bulletin no. 7: *Fire and the design of schools*. 4th edn., HMSO, London, 1971.
43. Department of Employment, *Means of escape in case of*

fire in offices, shops and railway premises, Health and safety at work, new series, no. 40, 2nd edn., HMSO, London, 1973.
44. Department of the Environment, The Building Regulations, 1972: Statutory instruments, 1972, no. 317, HMSO, London.
45. Department of the Environment, *Pedestrian safety*, HMSO, London, 1973.
46. Department of Health and Social Security, *Building for mentally handicapped people*, HMSO, London, 1971.
47. *Designing for the handicapped*, Bayes, K. and Franklin, Sandra (eds.), George Godwin, London, 1971.
48. Dudnik, E. E. and Krawczyk, R., An evaluation of space planning methodologies, in Preiser, W. F. E. (ed.), *Environmental design research*, vol. 1, Dowden, Hutchinson and Ross, Stroudsburg, Pennsylvania, 1973.
49. Fire Protection Association, *Fire prevention design guide*, Planning for fire safety in buildings, no. 4, London, 1969.
50. Fire Research Station, *Movement of smoke on escape routes in buildings*, proceedings of a symposium held 9 and 10 April, 1969, HMSO, London, 1971.
51. Fitch, J. M. *et al.*, The dimensions of stairs, *Scientific American*, **231**, part 4, 82–90, 1974.
52. Fleming, J., *Lift A1: a computer simulation for the design of lift installations in buildings*, University of Strathclyde, Abacus occasional paper no. 13.
53. Fletcher, P. T., The planning of lift installations in commercial buildings, *RIBA Journal*, **61**, 276–84, 1954.
54. Foot, N. I. S., Pedestrian traffic flows, *DMG. DRS Journal: design research and methods*, **7**, part 2, 162–7, 1973.
55. Forwood, B. S. and Gero, J. S., *Computer simulated lift design-analysis*, Department of Architectural Science, University of Sydney, Computer report CR9, 1970.
56. Frederking, M. A. and Penz, A. J., An integrated methodology for office building elevator design, in Preiser, W. F. E. (ed.), *Environmental design research*, vol. 1, Dowden, Hutchinson and Ross, Stroudsburg, Pennsylvania, 1973.

57. Fruin, J. J., *Designing for pedestrians: a level-of-service concept*, Highway Research Record no. 355, 1971.
58. Fruin, J. J., *Pedestrian planning and design*, New York, Metropolitan Association of Urban Designers and Environmental Planners, 1971.
59. Galbreath, M., Time of evacuation by stairs in high buildings, *Fire Fighting in Canada*, February, 1969, pp. 6–10.
60. Gaver, D. P. and Powell, B. A., Variability in round-trip times for an elevator car during up-peak, *Transportation Science*, **5**, part 2, 169–78, 1971.
61. Gero, J. S., *The application of operations research to architecture – a review*, Department of Architectural Science, University of Sydney, Computer report CR21, 1973.
62. Gilbert, L. J., Outline of the modern fluid-powered lift, *The Building Services Engineer*, **40**, A22–4, 1972.
63. Goldsmith, S., *Designing for the disabled*, 2nd edn., Royal Institute of British Architects, London, 1967.
64. Goldsmith, S., Mobility housing, *Architects' Journal*, **160**, 43–50, 1974.
65. Greater London Council, *Code of practice: means of escape in case of fire*, Greater London Council, London, 1974.
66. Grierson, R., *Electric lift equipment for modern building*, Chapman and Hall, London, 1923.
67. Hammond and Champness, Ltd., *The Hammond and Champness lift book*, London, 1956.
68. Hankin, B. D. and Wright, R. A., Passenger flow in subways, *Operational Research Quarterly*, **9**, part 2, 81–8, 1958.
69. Harding, J. D., Lifts in high rise buildings, *The Architect and Building News*, **223**, 783–4, 1963.
70. Heindrich, W. H., Fundamentals of elevator control, *Elevator World*, January 1963, 16–18, 31.
71. Hoel, L. A., Pedestrian travel rates in central business districts, *Traffic Engineering*, **38**, 10–13, 1968.
72. Home Office, Scottish Home and Health Department, *Guides to the Fire Precautions Act 1971, 1. Hotels and boarding houses*, HMSO, London, 1972.

73. Honey, L. W., *Lifts*, Marryat and Scott, London, 1946.
74. Hunt, S. T., Modern control system for groups of lifts, *Electrical Energy*, **3**, 18–26, 1959.
75. Hunt, S. T., Control of high speed lifts—a continuous pattern system, *GEC Journal of Science and Technology*, **32**, 117–24, 1965.
76. Hunt, S. T. and Bedford, R. J., Automatic control of groups of lifts. A new concept on supervisory systems, *G.E.C. Journal*, **31**, part 2, 84–91, 1964.
77. Hutton, B. P., Lifts, *Architects' Yearbook 9*, Elek, London, 1960, pp. 68–78.
78. Hutton, G. et al., Building investigation and planning: internal circulation, *Architects' Journal*, **137**, 589–602, 1963.
79. Jones, Bassett, The probable number of stops made by an elevator, *General Electric Review*, **26**, 583–7, 1923.
80. Jones, B. W., Simple analytical method for evaluating lift performance during an up-peak, *Transportation Research*, **5**, 301–7, 1971.
81. Jones, B. W., On building simulation models of lift systems, *Build International*, **6**, 225–43, 1973.
82. Julius, A. F., Lifts and escalators, *Architectural Science Review*, **2**, 9–22, 1959.
83. Kinsey, B. Y. and Sharp, H. M., *Environmental technologies in architecture*, Prentice-Hall, Englewood Cliffs, 1963.
84. Knight, T. L. and Duck, A. E., Cost of lifts, *Architects' Journal*, **135**, 361–6, 1962.
85. Langdon-Thomas, G. J., *Fire safety in buildings*, Adam and Charles Black, London, 1972.
86. Lawler, E. L., The quadratic assignment problem, *Management Science*, **9**, 586–99, 1963.
87. Lee, A. M., *Applied queueing theory*, Macmillan, London, 1966.
88. Leibowitz, M. A., An approximate method for treating a class of multiqueue problems, *IBM Journal*, **5**, 204–9, 1961.
89. Lerch, C. W., Basic factors in planning elevator systems, *Architectural Record*, **138**, 181–2, 1965.

90. Levin, P. H., Use of graphs to decide the optimum layout of buildings, *Architects' Journal*, **140**, 809–15, 1964.
91. Levy, D. *et al.*, Optimum elevator control using an on-line computer to divide building into equal-interval zones, in *Proceedings of the 6th international symposium on transportation, Sydney*, Edited by D. J. Buckley, Elsevier, New York, 1974.
92. Lindus, K. A., Vertical transportation, *Journal of the Institution of Heating and Ventilating Engineers*, **36**, 320–6, 1969.
93. Linzey, M. P. T., Optimum lift design for tall buildings, *Building Science*, **8**, 27–32, 1973.
94. London Transport Executive: (a) Research report no. 95, Second report of the operational research team on the capacity of footways, London, 1958; (b) O. R. Memorandum M258, The capacity of passageways for unidirectional and for crossing flows of pedestrians, London, 1972.
95. McGuinness, W. J. and Stein, B., *Mechanical and electrical equipment for buildings*, 5th edn., Wiley, New York, 1971.
96. Marchant, E. W., Escape routes, in Marchant, E. W. (ed.), *A complete guide to fire and buildings*, Medical and Technical Publishing, Lancaster, 1972, pp. 59–65.
97. Marmot, A. and Gero, J. S., Towards the development of an empirical model of elevator lobbies. *Building Science*, **9**, 277–88, 1974.
98. Menon, S. R., Elevator kinetics, *Elevator World*, April 1965, 18–23.
99. Ments, M. Van., Hospital planning: internal traffic in the general hospital, *Architects' Journal*, **138**, 27–30, 1963.
100. Ministry of Housing and Local Government, *Service cores in high flats*, Design bulletin no. 3, part 2, HMSO, London, 1962.
101. Ministry of Works, *Post-war building studies no. 9: mechanical installations*, HMSO, London, 1944.
102. Ministry of Works, *Post-war building studies no. 29: fire grading of buildings, Parts II, III and IV*, HMSO, London, 1952.

103. Morley, G., Installations, mechanical: lifts, escalators, paternosters, *Architects' Journal*, **136**, 233–58, 1962.
104. Morris, E. W., Lifts: the guiding factors for specification. *The Building Services Engineer*, **40**, A26–8, 1972.
105. Moseley, L., A rational design theory for planning buildings based on the analysis and solution of circulation problems, *Architects' Journal*, **138**, 525–37, 1963.
106. Moucka, J., Precision methods in practical planning, *Architektura, CSSR*, **2**, 128–32, 1963. Translated by H. Q. Panel, Building Research Station Library Communication no. 1270.
107. National Fire Protection Association, *Code for safety to life from fire in buildings and structures* (formerly *Building exits code*), NFPA no. 101, Washington, 1970.
108. Navin, F. P. D. and Wheeler, R. J., Pedestrian flow characteristics, *Traffic Engineering*, **19**, 30–6, 1969.
109. Nuder, A. and Johnsson, B., Hissar och trappor i kontorshus (in Swedish), *Byggmästaren*, **45**, 173–84, 1966.
110. Oeding, D., Verkehrsbelastung und Dimensionierung von Gehwegen und anderen Anlagen des Fussgängerverkehrs (Traffic load and dimensioning of footways and other constructions for pedestrian traffic) (in German), *Strassenbau und Strasserverkehrstechnik*, **22**, Bonn, 1963.
111. O'Flaherty, C. A. and Parkinson, M. H., Movement on a city centre footway, *Traffic engineering and control*, **13**, 434–8, 1972.
112. Older, S. J., Movement of pedestrians on footways in shopping streets, *Traffic engineering and control*, **10**, 160–3, 1968.
113. Oplinger, K. A. et al., A high-performance elevator control system, *Electrical Engineering*, **81**, 187–93, 1962.
114. Parnell, A. C. et al., How the Fire Precautions Act affects the architect, *Architects' Journal*, **155**, 1339–44, 1393–7, 1453–8; **156**, 37–40, 1972.
115. Peschel, R., Untersuchungen über die Leistangsfähigkeit ungeschützter Fussgängerüberwege (Research into the capacity of unprotected pedestrian crossings) (in German), *Strassertechnik*, **5**, part 6, 63–7, 1957.

116. Petigny, B., Les calculations des ascenseurs, *Transportation Research*, **6** (1), 19–38, 1972.
117. Phillips, R. J., Computerised approaches to circulation, *Building*, **216**, 16/117–122, 1969.
118. Phillips, R. S., *Electric lifts*, 6th edn., Pitman, London, 1973.
119. Pike, A., Lifts, escalators, pedestrian conveyors, *Architectural Design*, **37**, 294–6, 1967.
120. Port, L. W., The Port elevator system, *Architectural Science Review*, **11**, part 2, 52–7, 1968.
121. Powell, B. A., Optional elevator banking under heavy up-traffic, *Transportation Science*, **5**, part 2, 109–21, 1971.
122. Preiser, W. F. E., An analysis of unobtrusive observations of pedestrian movement and stationary behaviour in a shopping mall, in Küller, R. (ed.), *Architectural psychology: proceedings of the Lund Conference*, Dowden, Hutchinson and Ross, Stroudsburg, Pennsylvania, 1973, pp. 287–300.
123. Rawlinson, C. and Doidge, C., Dynamic space allocation, *Architectural Research and Teaching*, **1**, part 3, 4–10, 1971.
124. Reimer, K., Die Bewegung von menschen auf festen treppen (Movement of people on fixed stairs) (in German), *Glasers annalen*, **78**, part 2, 50–1, 1954.
125. Richardson, M., Physiological responses and energy expenditures of women using stairs of three designs, *Journal of Applied Physiology*, **21**, 1078–82, 1966.
126. Road Research Laboratory, *Research on road traffic*, HMSO, London, 1965.
127. Roberts, D. N., Electric and electro-hydraulic lifts and hoists, *Architects' Journal*, **158**, 211–15, 1973.
128. Ruchelman, L. I., Tall building decision-making and people: a case profile of the Pan Am building, Paper to the IABSE-ISE Joint Conference, St. Catherine's College, Oxford, 17–19 September, 1974.
129. Ruiz-Pala, E. *et al.*, *Waiting-line models*, Reinhold, New York, 1967.
130. Schay, G., Approximate methods for a multiqueueing problem, *IBM Journal*, **6**, 246–9, 1962.

131. Scottish Development Department, Building Standards (Scotland) Regulations, 1963. Explanatory memorandum, part 5: means of escape from fire and assistance to fire services; Statutory instruments 1963, no. 1896 (c. 17) (s. 101), HMSO, Edinburgh; The Building Standards (Scotland) (Consolidation) Regulations, 1971; Statutory instruments 1971, no. 2052 (s. 218), HMSO, Edinburgh.
132. Sharpe, R., Optimum space allocation within buildings, *Building Science*, **8**, 201–5, 1973.
133. Shirley, E., Circulation analysis and lift placing in hospitals, *Computer-aided Design*, **6**, 206–10, 1974.
134. Simmons, D. M., One-dimensional space allocation: an ordering algorithm, *Operations Research*, **17**, 812–26, 1969.
135. Steadman, P. and March, L., *The geometry of the environment: an introduction to spatial organization in design*, Royal Institute of British Architects, London, 1971.
136. Stilitz, I. B., The role of static pedestrian groups in crowded spaces, *Ergonomics*, **12**, part 6, 821–39, 1969.
137. Strakosch, G. R., How to plan elevator service for college buildings, *College and University Business*, January and February, 1964.
138. Strakosch, G. R., *Vertical transportation: elevators and escalators*, Wiley, New York, 1967.
139. Surti, V. H. and Burke, T. J., Investigation of the capacity of the White House sidewalk for orderly demonstrations, *Highway Research Record*, no. 355, 1971.
140. Swartz, W. W., Optimizing space requirements for elevators, *Architectural Record*, **147**, 133–6, 1970.
141. Tabor, P., *Pedestrian circulation in offices*, working paper no. 17; *Systematic activity location*, working paper no. 18; *Analysis of communication patterns*, working paper no. 19; *Evaluation of routes*, working paper no. 20; Land Use and Built Form Studies, University of Cambridge School of Architecture, 1969.
142. Tocher, K. D., *The art of simulation*, English Universities Press, London, 1963.

143. Togawa, K., *Study on fire escapes basing on the observation of multitude currents* (in Japanese), Report No. 14, Building Research Institute, Japan, 1955.
144. Tough, J. M. and O'Flaherty, C. A., *Passenger conveyors*, Ian Allan, London, 1971.
145. Tregenza, P. R., Association between building height and cost, *Architects' Journal*, **156**, 1031–2, 1972.
146. Tregenza, P. R., The prediction of passenger lift performance, *Architectural Science Review*, **15**, 49–54, 1972.
147. Tregenza, P. R., Observations of pedestrian movement (unpublished report), University of Nottingham, 1974.
148. Tregenza, P. R., The theoretical capacity of paternoster lifts, *Building Science*, **9**, 79–83, 1974.
149. Turner, F. S. P., Preliminary planning for a new tube railway across London, *Proc. Institution of Civil Engineers*, **12**, 19–38, 1959.
150. Underdown, G. W., *Practical fire precautions* (Chapter 2, Means of escape from fire), Gower Press, London, 1971, pp. 14–30.
151. University Facilities Research Center, *Horizontal and vertical circulation in university instructional and research buildings*, University of Wisconsin, 1961 (3rd printing, 1964).
152. University of Edinburgh, *Planning for disabled people in the urban environment*, Central Council for the Disabled, London, 1969.
153. Vartanov, G. L. and Ponkratov, B. K., The determination of design group loads of lift installations by probability modelling methods (in Russian. Translation RTS 8729, British Library Lending Division). *Elektrichestvo*, **10**, 88–90, 1970.
154. Voort, A. P. F. van der, Keuze van de liftinstallatie bij woongebouwen (in Dutch), *Bouw*, **23**, 252–8, 1968.
155. Wadsworth, Wm. and Sons, Ltd., *The Wadsworth book of traffic studies*, Bolton, 1973.
156. Walter, F., *Four architectural movement studies*, Disabled Living Foundation, London, 1971.
157. Ward, C. (ed.), *Vandalism*, Architectural Press, London, 1973.

158. Ward, J. S. and Randall, P., Optimum dimensions for domestic stairways: a preliminary study, *Architects' Journal*, **146**, 29–34, 1967.
159. Warren, A., A note on lift waiting times and failures; Appendix E in Jephcott, A. P. and Robinson, H., *Homes in high flats*. University of Glasgow social and economic studies: occasional papers 13, Oliver and Boyd, Edinburgh, 1971, pp. 159–64.
160. Weinberger, W., Berechnung von Förderleistung und Wartezeiten bei Aufzugsanlagen (Calculating the capacity and waiting time of lift installations) (in German), *Förden und Heben*, **18**, part 2, 105–9, 1968.
161. Whitehead, B. and Eldars, M. Z., An approach to the optimum layout of single-storey buildings. *Architects' Journal*, **139**, 1373–80, 1964.
162. Williams, F. H., Selection of passenger lifts for office buildings, *Architects' Journal*, **156**, 331–2, 1972.
163. Williams, F. H., Selection of passenger lifts for tall office buildings, *Architects' Journal*, **160**, 1337–40, 1974.
164. Williamson, E. and Bretherton, M. H., *Tables of the negative binomial probability distribution*, Wiley, London, 1963.
165. Whyman, P., Movement of pupils in comprehensive schools, *Building*, **216**, 20/86–90, 1969.
166. Wurst, H., Zur Dimensionierung von Verkehrswegen in Gebäuden, *Der Aufbau*, **28**, 53–6, 1973.

APPENDIX 1

Table and graphs for selecting lifts

Calculations of lift performance are sensitive to small variations in operating times. The parameters given in Table 1 are those on which the graphs in this appendix were calculated. They may be adopted for other computations, using the tables of Appendix 2, but when an accurate prediction of the perfor-

Table 1*   Typical lift parameters and passenger transfer times

	t_h	t_{s1}	t_{s2}	t_{ss}
0·75 m/s lift, geared a.c. motor, 800-mm doors, non-premature opening	4·40	9·50	9·50	9·50
1·0 m/s lift, geared a.c. motor, 800-mm doors, non-premature opening	3·30	10·50	10·50	10·50
1·5 m/s lift, geared v–v motor, 1100-mm doors with premature opening	2·20	6·80	6·80	6·80
2·5 m/s lift, gearless v–v motor, 1100-mm doors with premature opening	1·32	7·00	6·90	6·80
3·5 m/s lift, gearless v–v motor, 1100-mm doors with premature opening	0·94	7·40	7·60	7·50
5·0 m/s lift, gearless v–v motor, 1100-mm doors with premature opening	0·66	7·60	8·20	8·30

* The symbols used are defined in Section 2.2, p. 23.

mance of a particular lift system is required, specific data should be obtained from the manufacturer of the equipment.

The values in Table 1 apply to a storey height of 3·3 m. The figures for t_{s1}, t_{s2} and t_{ss} do not change greatly with variation of storey height. The value of E_h in other cases is given by

$$E_h = d/v$$

where d is the storey height and v is the maximum speed of the lift.

Passenger transfer times are affected by the depth of the car, the door opening width, the lift lobby design and the total number of people using the lift. Considerable variation is found between different installations and there are insufficient survey data to quantify all the relevant factors. To calculate the following graphs these figures were assumed:

$t_p = 2\cdot 1$ s (unidirectional traffic)
$t_p = 3\cdot 0$ s (random inter-floor traffic)

Phillips [118] gives graphs that relate entry and exit times to the total number of passengers and the number of stops with unidirectional traffic.

Table 2. Nominal maximum car capacities

Load (kg)	Capacity (P)
450	6
600	8
750	10
900	12
1200	16
1500	20
1800	24
2100	28

134 The Design of Interior Circulation

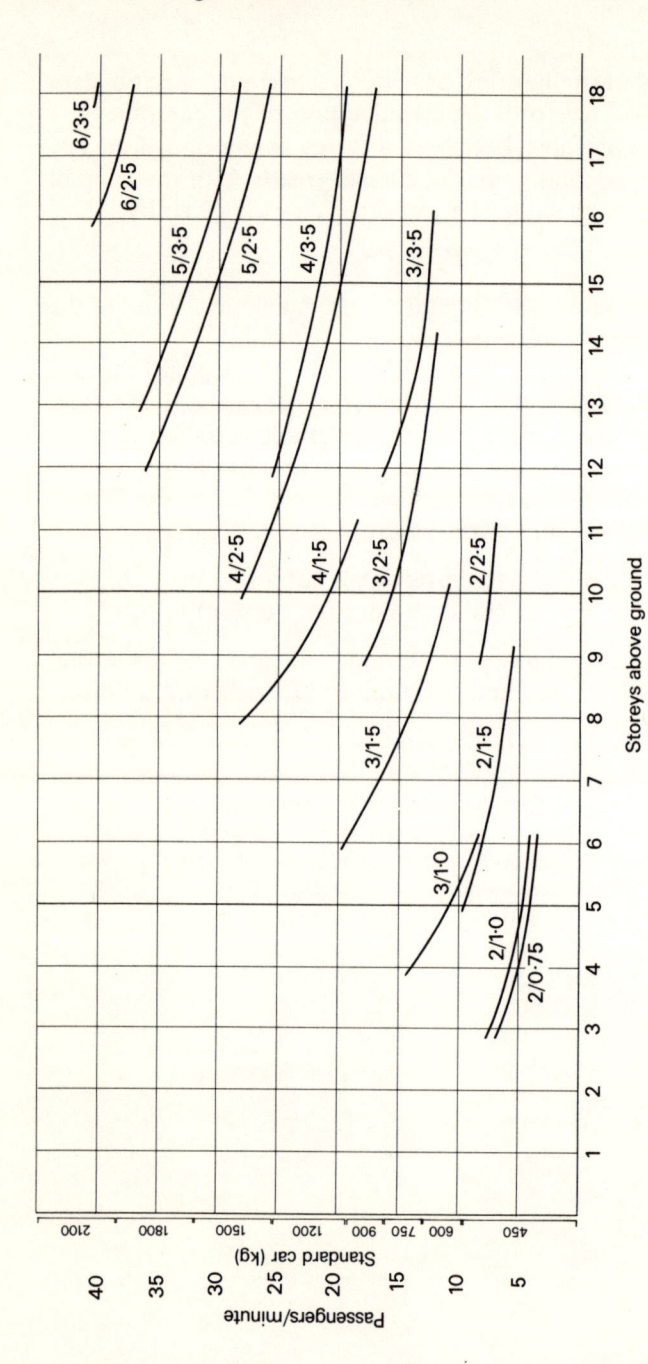

A. Lift systems with a mean interval of 30 s. Unidirectional traffic from ground floor. 3·3 m storey height (Note: In Figs. A–D, 2/0·75 indicates a system of two lifts of contract speed 0·75 m/s with the characteristics given in Table 1 of Appendix 1, and analogously for other values)

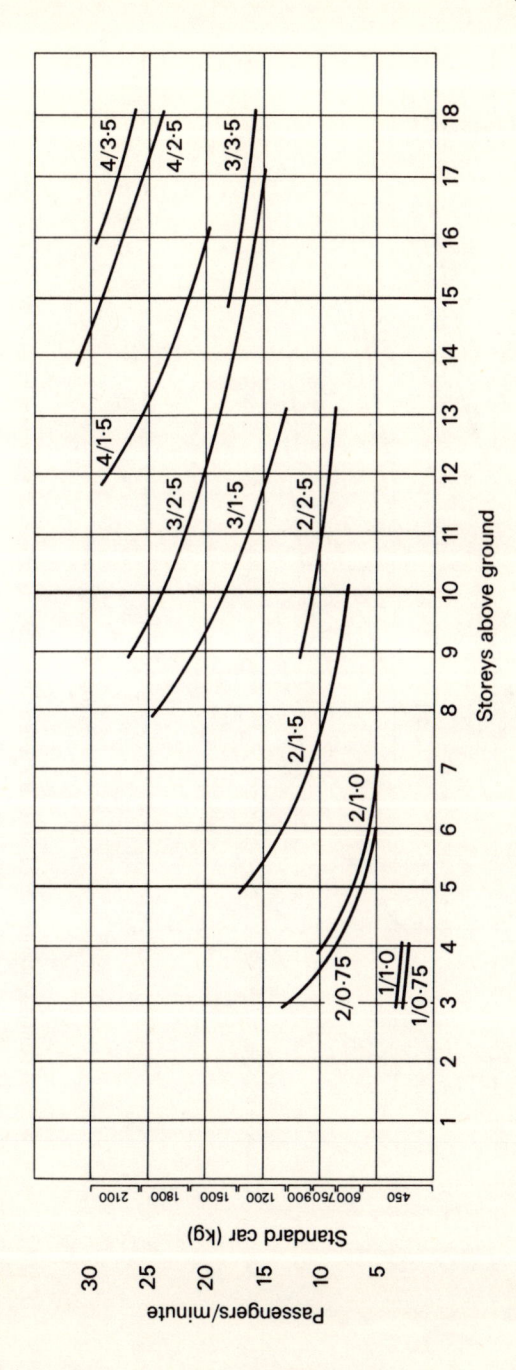

B. Lift systems with a mean interval of 45 s. Unidirectional traffic from the ground floor. 3·3 m storey height

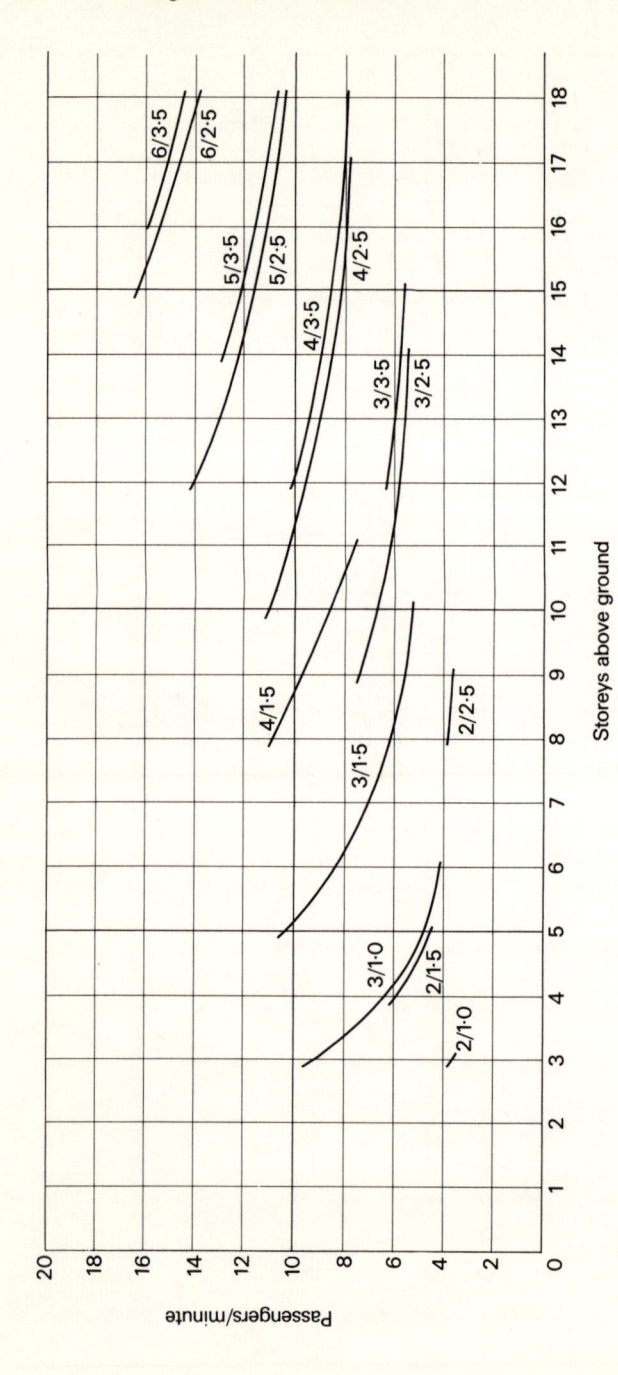

C Lift systems with an equivalent interval of 30 s. Random inter-floor traffic. 3·3 m storey height. With uniform distribution of traffic 450 kg capacity cars would be adequate

Appendix 1 **137**

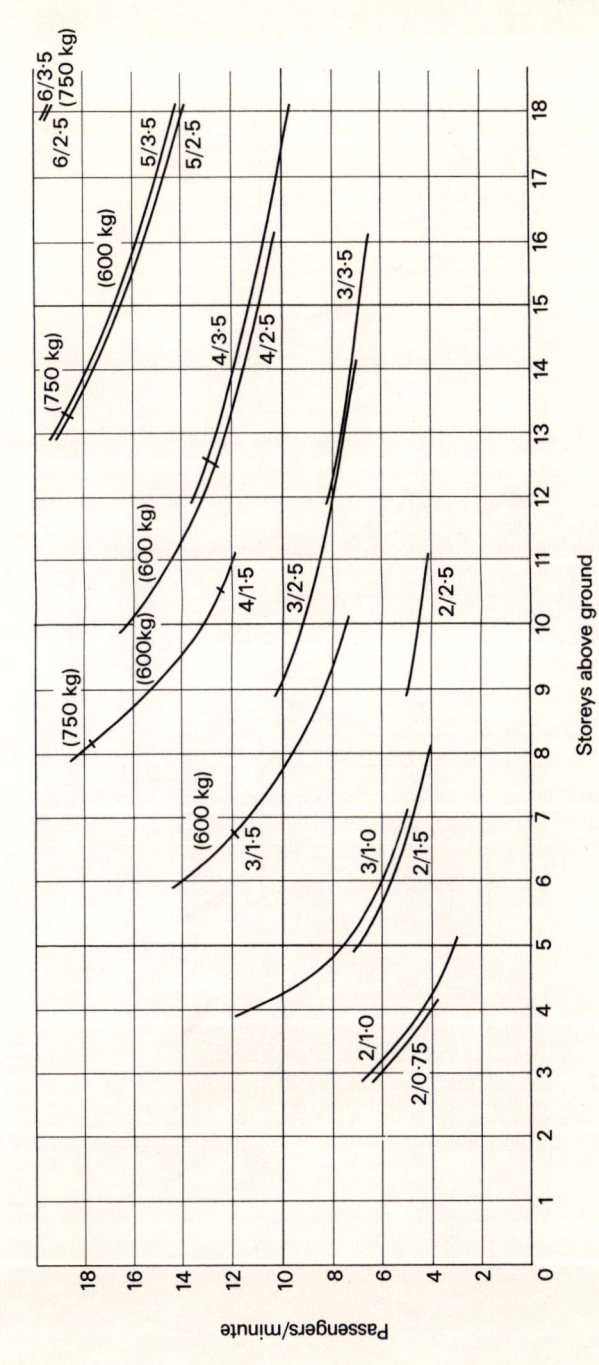

D. Lift systems with an equivalent interval of 45 s. Random inter-floor traffic. With uniform distribution of traffic 450 kg cars would be adequate except where marked

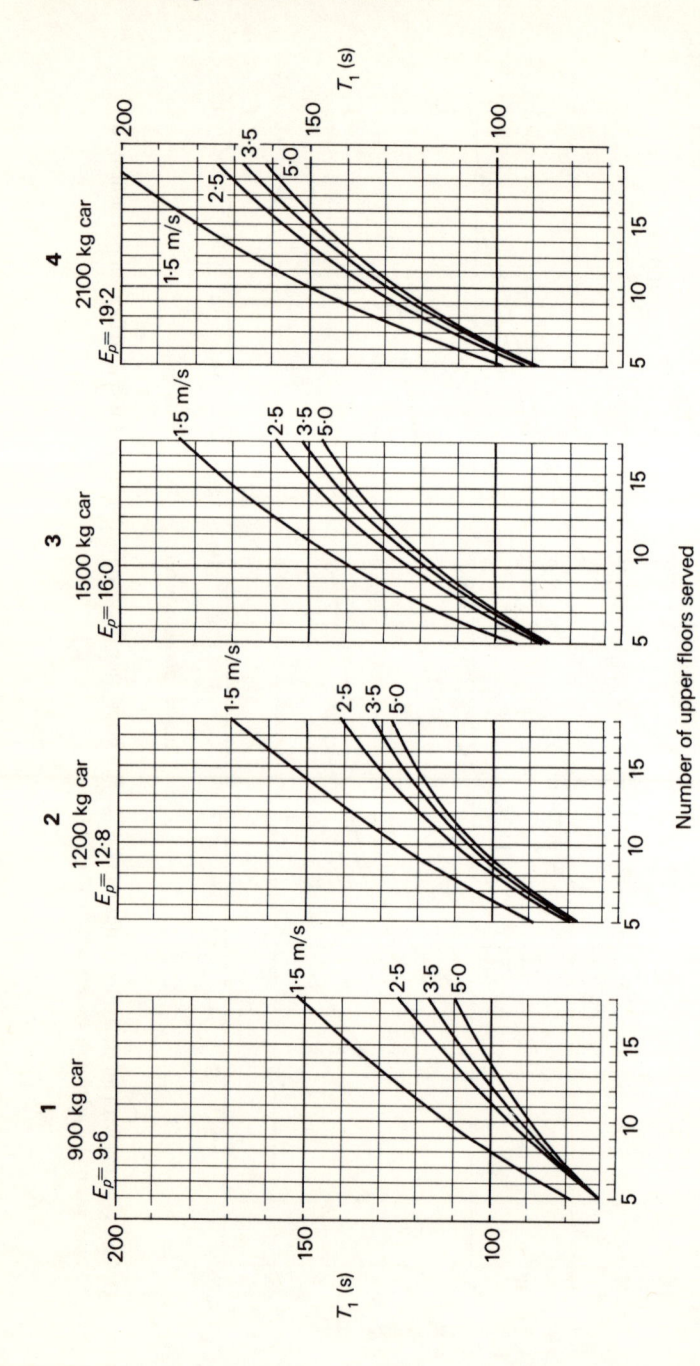

E. Cycle times of single lifts with standard cars at 80 per cent maximum load. 3·3 m storey height

APPENDIX 2

Tables of E_h, E_{s1}, E_{s2}, E_{ss} and E_p for unidirectional traffic and random inter-floor traffic

Unidirectional traffic

Floors above ground: 3

p_0	E_h	E_{s1}	E_{s2}	E_{ss}	E_p
0·050	2·9	2·8	0·1	1·0	9·0
0·100	2·9	2·5	0·3	0·9	6·9
0·150	2·8	2·3	0·4	0·9	5·7
0·200	2·8	2·1	0·4	0·8	4·8
0·250	2·7	1·9	0·5	0·8	4·2
0·300	2·6	1·7	0·6	0·8	3·6
0·350	2·5	1·6	0·6	0·7	3·1
0·400	2·4	1·4	0·6	0·7	2·7
0·450	2·3	1·3	0·6	0·7	2·4
0·500	2·1	1·1	0·6	0·6	2·1
0·550	2·0	1·0	0·6	0·6	1·8

Unidirectional traffic
Floors above ground: 4

p_0	E_h	E_{s1}	E_{s2}	E_{ss}	E_p
0·050	3·9	3·7	0·1	1·0	12·0
0·100	3·9	3·3	0·3	1·0	9·2
0·150	3·8	3·0	0·4	1·0	7·6
0·200	3·8	2·7	0·4	1·0	6·4
0·250	3·7	2·4	0·5	1·0	5·5
0·300	3·6	2·2	0·6	1·0	4·8
0·350	3·5	1·9	0·6	1·0	4·2
0·400	3·4	1·7	0·6	1·0	3·7
0·450	3·2	1·5	0·6	1·0	3·2
0·500	3·1	1·3	0·6	1·0	2·8
0·550	2·9	1·1	0·6	1·0	2·4
0·600	2·7	1·0	0·6	0·9	2·0
0·650	2·5	0·8	0·5	0·9	1·7

Floors above ground: 5

p_0	E_h	E_{s1}	E_{s2}	E_{ss}	E_p
0·050	4·9	4·6	0·2	1·0	15·0
0·100	4·9	4·1	0·3	1·0	11·5
0·150	4·8	3·7	0·5	1·1	9·5
0·200	4·8	3·4	0·6	1·1	8·0
0·250	4·7	3·0	0·6	1·1	6·9
0·300	4·6	2·7	0·7	1·2	6·0
0·350	4·5	2·3	0·7	1·2	5·2
0·400	4·3	2·1	0·7	1·2	4·6
0·450	4·2	1·8	0·7	1·2	4·0
0·500	4·0	1·5	0·7	1·3	3·5
0·550	3·8	1·3	0·7	1·2	3·0
0·600	3·6	1·1	0·6	1·2	2·6
0·650	3·4	0·9	0·6	1·2	2·2
0·700	3·1	0·7	0·5	1·1	1·8

Unidirectional traffic
Floors above ground: 6

p_0	E_h	E_{s1}	E_{s2}	E_{ss}	E_p
0·050	5·9	5·5	0·2	1·0	18·0
0·100	5·9	5·0	0·4	1·0	13·8
0·150	5·8	4·5	0·6	1·1	11·4
0·200	5·8	4·0	0·7	1·1	9·7
0·250	5·7	3·6	0·8	1·2	8·3
0·300	5·6	3·2	0·8	1·2	7·2
0·350	5·5	2·8	0·8	1·3	6·3
0·400	5·3	2·4	0·8	1·4	5·5
0·450	5·2	2·1	0·8	1·4	4·8
0·500	5·0	1·8	0·8	1·4	4·2
0·550	4·8	1·5	0·7	1·5	3·6
0·600	4·6	1·2	0·7	1·4	3·1
0·650	4·3	1·0	0·6	1·4	2·6
0·700	3·9	0·8	0·5	1·3	2·1
0·750	3·5	0·6	0·5	1·2	1·7

Floors above ground: 7

p_0	E_h	E_{s1}	E_{s2}	E_{ss}	E_p
0·050	6·9	6·4	0·3	1·0	21·0
0·100	6·9	5·8	0·5	1·0	16·1
0·150	6·8	5·2	0·7	1·1	13·3
0·200	6·8	4·6	0·8	1·2	11·3
0·250	6·7	4·1	0·9	1·2	9·7
0·300	6·6	3·6	0·9	1·3	8·4
0·350	6·5	3·2	1·0	1·4	7·3
0·400	6·3	2·8	1·0	1·5	6·4
0·450	6·2	2·4	0·9	1·5	5·6
0·500	6·0	2·0	0·9	1·6	4·9
0·550	5·8	1·7	0·8	1·6	4·2
0·600	5·5	1·4	0·8	1·6	3·6
0·650	5·2	1·1	0·7	1·6	3·0
0·700	4·9	0·9	0·6	1·6	2·5
0·750	4·4	0·7	0·5	1·5	2·0

Unidirectional traffic
Floors above ground: 8

p_0	E_h	E_{s1}	E_{s2}	E_{ss}	E_p
0·050	7·9	7·3	0·3	1·0	24·0
0·100	7·9	6·6	0·6	1·1	18·4
0·150	7·8	5·9	0·8	1·1	15·2
0·200	7·8	5·3	0·9	1·2	12·9
0·250	7·7	4·7	1·0	1·3	11·1
0·300	7·6	4·1	1·1	1·4	9·6
0·350	7·5	3·6	1·1	1·5	8·4
0·400	7·3	3·1	1·1	1·6	7·3
0·450	7·2	2·7	1·1	1·7	6·4
0·500	7·0	2·3	1·0	1·7	5·5
0·550	6·8	1·9	0·9	1·8	4·8
0·600	6·5	1·5	0·8	1·8	4·1
0·650	6·2	1·2	0·7	1·8	3·4
0·700	5·8	1·0	0·6	1·8	2·9
0·750	5·3	0·7	0·5	1·7	2·3
0·800	4·7	0·5	0·4	1·5	1·8

Floors above ground: 9

p_0	E_h	E_{s1}	E_{s2}	E_{ss}	E_p
0·050	8·9	8·2	0·4	1·0	27·0
0·100	8·9	7·4	0·7	1·1	20·7
0·150	8·8	6·6	0·9	1·1	17·1
0·200	8·8	5·9	1·1	1·2	14·5
0·250	8·7	5·3	1·2	1·3	12·5
0·300	8·6	4·6	1·2	1·4	10·8
0·350	8·5	4·0	1·3	1·6	9·4
0·400	8·3	3·5	1·2	1·7	8·2
0·450	8·2	3·0	1·2	1·8	7·2
0·500	8·0	2·5	1·1	1·9	6·2
0·550	7·8	2·1	1·0	1·9	5·4
0·600	7·5	1·7	0·9	2·0	4·6
0·650	7·2	1·3	0·8	2·0	3·9
0·700	6·8	1·0	0·7	1·9	3·2
0·750	6·2	0·8	0·5	1·9	2·6
0·800	5·5	0·6	0·4	1·7	2·0

Unidirectional traffic

Floors above ground: 10

p_0	E_h	E_{s1}	E_{s2}	E_{ss}	E_p
0·050	9·9	9·1	0·4	1·0	30·0
0·100	9·9	8·2	0·7	1·1	23·0
0·150	9·8	7·4	1·0	1·2	19·0
0·200	9·8	6·6	1·2	1·3	16·1
0·250	9·7	5·8	1·3	1·4	13·9
0·300	9·6	5·1	1·4	1·5	12·0
0·350	9·5	4·5	1·4	1·6	10·5
0·400	9·3	3·8	1·4	1·8	9·2
0·450	9·2	3·3	1·3	1·9	8·0
0·500	9·0	2·8	1·3	2·0	6·9
0·550	8·8	2·3	1·1	2·1	6·0
0·600	8·5	1·8	1·0	2·1	5·1
0·650	8·2	1·5	0·9	2·2	4·3
0·700	7·7	1·1	0·7	2·1	3·6
0·750	7·2	0·8	0·6	2·0	2·9
0·800	6·4	0·6	0·4	1·9	2·2
0·850	5·4	0·4	0·3	1·6	1·6

Floors above ground: 11

p_0	E_h	E_{s1}	E_{s2}	E_{ss}	E_p
0·050	10·9	10·0	0·5	1·0	33·0
0·100	10·9	9·0	0·8	1·1	25·3
0·150	10·8	8·1	1·1	1·2	20·9
0·200	10·8	7·2	1·3	1·3	17·7
0·250	10·7	6·4	1·5	1·4	15·2
0·300	10·6	5·6	1·5	1·6	13·2
0·350	10·5	4·9	1·6	1·7	11·5
0·400	10·3	4·2	1·5	1·9	10·1
0·450	10·2	3·6	1·5	2·0	8·8
0·500	10·0	3·0	1·4	2·1	7·6
0·550	9·8	2·5	1·3	2·2	6·6
0·600	9·5	2·0	1·1	2·3	5·6
0·650	9·2	1·6	1·0	2·3	4·7
0·700	8·7	1·2	0·8	2·3	3·9
0·750	8·1	0·9	0·6	2·2	3·2
0·800	7·3	0·6	0·5	2·0	2·5
0·850	6·3	0·4	0·3	1·7	1·8

Unidirectional traffic

Floors above ground: 12

p_0	E_h	E_{s1}	E_{s2}	E_{ss}	E_p
0·050	11·9	10·9	0·5	1·0	35·9
0·100	11·9	9·8	0·9	1·1	27·6
0·150	11·8	8·8	1·2	1·2	22·8
0·200	11·8	7·8	1·4	1·3	19·3
0·250	11·7	6·9	1·6	1·5	16·6
0·300	11·6	6·1	1·7	1·6	14·4
0·350	11·5	5·3	1·7	1·8	12·6
0·400	11·3	4·6	1·7	2·0	11·0
0·450	11·2	3·9	1·6	2·1	9·6
0·500	11·0	3·3	1·5	2·2	8·3
0·550	10·8	2·7	1·4	2·4	7·2
0·600	10·5	2·2	1·2	2·4	6·1
0·650	10·2	1·7	1·0	2·5	5·2
0·700	9·7	1·3	0·8	2·4	4·3
0·750	9·1	0·9	0·7	2·3	3·5
0·800	8·3	0·7	0·5	2·2	2·7
0·850	7·1	0·4	0·3	1·9	2·0

Floors above ground: 13

p_0	E_h	E_{s1}	E_{s2}	E_{ss}	E_p
0·050	12·9	11·8	0·5	1·0	38·9
0·100	12·9	10·6	1·0	1·1	29·9
0·150	12·8	9·5	1·3	1·2	24·7
0·200	12·8	8·5	1·6	1·4	20·9
0·250	12·7	7·5	1·7	1·5	18·0
0·300	12·6	6·6	1·8	1·7	15·7
0·350	12·5	5·7	1·9	1·9	13·6
0·400	12·3	4·9	1·8	2·1	11·9
0·450	12·2	4·2	1·7	2·2	10·4
0·500	12·0	3·5	1·6	2·4	9·0
0·550	11·8	2·9	1·5	2·5	7·8
0·600	11·5	2·3	1·3	2·6	6·6
0·650	11·1	1·8	1·1	2·6	5·6
0·700	10·7	1·4	0·9	2·6	4·6
0·750	10·1	1·0	0·7	2·5	3·7
0·800	9·2	0·7	0·5	2·3	2·9
0·850	8·0	0·4	0·4	2·0	2·1

Unidirectional traffic
Floors above ground: 14

p_0	E_h	E_{s1}	E_{s2}	E_{ss}	E_p
0·100	13·9	11·4	1·1	1·1	32·2
0·150	13·8	10·2	1·4	1·2	26·6
0·200	13·8	9·1	1·7	1·4	22·5
0·250	13·7	8·1	1·9	1·6	19·4
0·300	13·6	7·1	2·0	1·8	16·9
0·350	13·5	6·1	2·0	2·0	14·7
0·400	13·3	5·3	2·0	2·2	12·8
0·450	13·2	4·5	1·9	2·3	11·2
0·500	13·0	3·8	1·8	2·5	9·7
0·550	12·8	3·1	1·6	2·6	8·4
0·600	12·5	2·5	1·4	2·7	7·2
0·650	12·1	1·9	1·2	2·8	6·0
0·700	11·7	1·5	1·0	2·8	5·0
0·750	11·1	1·1	0·8	2·7	4·0
0·800	10·2	0·7	0·6	2·5	3·1
0·850	8·9	0·5	0·4	2·2	2·3

Floors above ground: 15

p_0	E_h	E_{s1}	E_{s2}	E_{ss}	E_p
0·100	14·9	12·2	1·1	1·1	34·5
0·150	14·8	11·0	1·5	1·2	28·5
0·200	14·8	9·8	1·8	1·4	24·1
0·250	14·7	8·6	2·0	1·6	20·8
0·300	14·6	7·6	2·1	1·8	18·1
0·350	14·5	6·6	2·1	2·0	15·7
0·400	14·3	5·6	2·1	2·2	13·7
0·450	14·2	4·8	2·0	2·4	12·0
0·500	14·0	4·0	1·9	2·6	10·4
0·550	13·8	3·3	1·7	2·8	9·0
0·600	13·5	2·6	1·5	2·9	7·7
0·650	13·1	2·1	1·3	2·9	6·5
0·700	12·7	1·6	1·0	2·9	5·4
0·750	12·0	1·1	0·8	2·8	4·3
0·800	11·1	0·8	0·6	2·6	3·3
0·850	9·8	0·5	0·4	2·3	2·4

Unidirectional traffic

Floors above ground: 16

p_0	E_h	E_{s1}	E_{s2}	E_{ss}	E_p
0·100	15·9	13·1	1·2	1·1	36·8
0·150	15·8	11·7	1·6	1·3	30·4
0·200	15·8	10·4	2·0	1·4	25·8
0·250	15·7	9·2	2·2	1·7	22·2
0·300	15·6	8·1	2·3	1·9	19·3
0·350	15·5	7·0	2·3	2·1	16·8
0·400	15·3	6·0	2·3	2·3	14·7
0·450	15·2	5·1	2·2	2·6	12·8
0·500	15·0	4·3	2·0	2·7	11·1
0·550	14·8	3·5	1·8	2·9	9·6
0·600	14·5	2·8	1·6	3·0	8·2
0·650	14·1	2·2	1·3	3·1	6·9
0·700	13·7	1·7	1·1	3·1	5·7
0·750	13·0	1·2	0·8	3·0	4·6
0·800	12·1	0·8	0·6	2·7	3·6
0·850	10·8	0·5	0·4	2·4	2·6
0·900	8·7	0·3	0·2	1·9	1·7

Floors above ground: 17

p_0	E_h	E_{s1}	E_{s2}	E_{ss}	E_p
0·100	16·9	13·9	1·3	1·1	39·1
0·150	16·8	12·4	1·8	1·3	32·3
0·200	16·8	11·0	2·1	1·5	27·4
0·250	16·7	9·8	2·3	1·7	23·6
0·300	16·6	8·5	2·4	1·9	20·5
0·350	16·5	7·4	2·4	2·2	17·8
0·400	16·3	6·4	2·4	2·4	15·6
0·450	16·2	5·4	2·3	2·7	13·6
0·500	16·0	4·5	2·1	2·9	11·8
0·550	15·8	3·7	1·9	3·0	10·2
0·600	15·5	3·0	1·7	3·2	8·7
0·650	15·1	2·3	1·4	3·2	7·3
0·700	14·7	1·7	1·2	3·2	6·1
0·750	14·0	1·3	0·9	3·1	4·9
0·800	13·1	0·8	0·6	2·9	3·8
0·850	11·7	0·5	0·4	2·5	2·8
0·900	9·5	0·3	0·2	2·0	1·8

Unidirectional traffic

Floors above ground: 18

p_0	E_h	E_{s1}	E_{s2}	E_{ss}	E_p
0·150	17·8	13·1	1·9	1·3	34·1
0·200	17·8	11·7	2·2	1·5	29·0
0·250	17·7	10·3	2·4	1·7	25·0
0·300	17·6	9·0	2·6	2·0	21·7
0·350	17·5	7·8	2·6	2·3	18·9
0·400	17·3	6·7	2·5	2·5	16·5
0·450	17·2	5·7	2·4	2·8	14·4
0·500	17·0	4·8	2·3	3·0	12·5
0·550	16·8	3·9	2·0	3·2	10·8
0·600	16·5	3·1	1·8	3·3	9·2
0·650	16·1	2·4	1·5	3·4	7·8
0·700	15·7	1·8	1·2	3·3	6·4
0·750	15·0	1·3	0·9	3·2	5·2
0·800	14·1	0·9	0·7	3·0	4·0
0·850	12·6	0·5	0·4	2·7	2·9
0·900	10·4	0·3	0·3	2·1	1·9

Inter-floor traffic

Floors above ground: 3

p_0	E_h	E_{s1}	E_{s2}	E_{ss}	E_p
0·050	3·0	6·0	0·0	0·0	35·9
0·100	3·0	6·0	0·0	0·0	27·6
0·150	3·0	6·0	0·0	0·0	22·8
0·200	3·0	6·0	0·0	0·0	19·3
0·250	3·0	5·9	0·0	0·0	16·6
0·300	3·0	5·9	0·0	0·0	14·4
0·350	3·0	5·8	0·1	0·0	12·6
0·400	3·0	5·7	0·1	0·0	11·0
0·450	3·0	5·6	0·2	0·0	9·6
0·500	3·0	5·4	0·2	0·0	8·3
0·550	3·0	5·2	0·3	0·0	7·2
0·600	2·9	4·9	0·4	0·1	6·1
0·650	2·9	4·5	0·5	0·1	5·2
0·700	2·8	4·0	0·6	0·2	4·3
0·750	2·7	3·5	0·6	0·2	3·5
0·800	2·6	3·0	0·7	0·3	2·7

Floors above ground: 4

p_0	E_h	E_{s1}	E_{s2}	E_{ss}	E_p
0·150	4·0	8·0	0·0	0·0	37·9
0·200	4·0	8·0	0·0	0·0	32·2
0·250	4·0	8·0	0·0	0·0	27·7
0·300	4·0	8·0	0·0	0·0	24·1
0·350	4·0	7·9	0·0	0·0	21·0
0·400	4·0	7·8	0·1	0·0	18·3
0·450	4·0	7·7	0·1	0·0	16·0
0·500	4·0	7·6	0·2	0·0	13·9
0·550	4·0	7·4	0·3	0·0	12·0
0·600	4·0	7·1	0·3	0·1	10·2
0·650	4·0	6·7	0·5	0·1	8·6
0·700	3·9	6·2	0·6	0·1	7·1
0·750	3·9	5·5	0·8	0·2	5·8
0·800	3·8	4·7	0·9	0·4	4·5
0·820	3·7	4·4	0·9	0·4	4·0
0·840	3·6	3·8	1·0	0·5	3·5
0·860	3·5	3·3	0·9	0·6	3·0

Inter-floor traffic

Floors above ground: 5

p_0	E_h	E_{s1}	E_{s2}	E_{ss}	E_p
0·300	5·0	10·0	0·0	0·0	36·1
0·350	5·0	10·0	0·0	0·0	31·5
0·400	5·0	9·9	0·0	0·0	27·5
0·450	5·0	9·9	0·1	0·0	24·0
0·500	5·0	9·8	0·1	0·0	20·8
0·550	5·0	9·6	0·2	0·0	17·9
0·600	5·0	9·3	0·3	0·0	15·3
0·650	5·0	8·9	0·4	0·1	12·9
0·700	5·0	8·4	0·6	0·1	10·7
0·750	4·9	7·7	0·8	0·2	8·6
0·800	4·9	6·7	1·0	0·3	6·7
0·820	4·8	6·2	1·1	0·4	6·0
0·840	4·8	5·6	1·1	0·5	5·2
0·860	4·7	5·0	1·2	0·6	4·5
0·880	4·5	4·3	1·2	0·7	3·8
0·900	4·3	3·6	1·1	0·8	3·2
0·910	4·2	3·3	1·1	0·9	2·8

Floors above ground: 6

p_0	E_h	E_{s1}	E_{s2}	E_{ss}	E_p
0·400	6·0	12·0	0·0	0·0	38·5
0·450	6·0	11·9	0·0	0·0	33·5
0·500	6·0	11·8	0·1	0·0	29·1
0·550	6·0	11·7	0·1	0·0	25·1
0·600	6·0	11·5	0·2	0·0	21·5
0·650	6·0	11·1	0·4	0·0	18·1
0·700	6·0	10·7	0·5	0·1	15·0
0·750	6·0	9·9	0·8	0·1	12·1
0·800	5·9	8·8	1·0	0·3	9·4
0·820	5·9	8·2	1·2	0·3	8·3
0·840	5·8	7·6	1·3	0·4	7·3
0·860	5·8	6·9	1·3	0·6	6·3
0·880	5·7	6·0	1·5	0·7	5·4
0·900	5·5	5·1	1·4	0·9	4·4
0·910	5·4	4·5	1·4	1·0	4·0
0·920	5·3	4·0	1·4	1·1	3·5
0·930	5·1	3·5	1·3	1·1	3·0

Inter-floor traffic

Floors above ground: 7

p_0	E_h	E_{s1}	E_{s2}	E_{ss}	E_p
0·500	7·0	13·9	0·1	0·0	38·8
0·550	7·0	13·8	0·1	0·0	33·5
0·600	7·0	13·7	0·1	0·0	28·6
0·650	7·0	13·4	0·3	0·0	24·1
0·700	7·0	12·9	0·5	0·0	20·0
0·750	7·0	12·2	0·7	0·1	16·1
0·800	7·0	11·0	1·1	0·2	12·5
0·820	6·9	10·4	1·2	0·3	11·1
0·840	6·9	9·8	1·4	0·4	9·8
0·860	6·9	8·8	1·5	0·5	8·4
0·880	6·8	7·8	1·6	0·7	7·2
0·900	6·7	6·6	1·7	0·9	5·9
0·910	6·6	6·2	1·7	1·0	5·3
0·920	6·5	5·6	1·7	1·1	4·7
0·930	6·3	4·8	1·7	1·3	4·1
0·940	6·1	4·0	1·6	1·4	3·5
0·950	5·8	3·2	1·4	1·5	2·9

Floors above ground: 8

p_0	E_h	E_{s1}	E_{s2}	E_{ss}	E_p
0·600	8·0	15·8	0·1	0·0	36·8
0·650	8·0	15·5	0·2	0·0	31·0
0·700	8·0	15·1	0·4	0·0	25·7
0·750	8·0	14·6	0·6	0·1	20·7
0·800	8·0	13·2	1·0	0·2	16·1
0·820	8·0	12·6	1·2	0·3	14·3
0·840	7·9	11·9	1·4	0·4	12·6
0·860	7·9	10·9	1·6	0·5	10·9
0·880	7·8	9·9	1·8	0·7	9·2
0·900	7·8	8·6	1·9	0·9	7·6
0·910	7·7	7·7	2·0	1·0	6·8
0·920	7·6	6·9	2·0	1·2	6·0
0·930	7·5	6·0	1·9	1·3	5·2
0·940	7·3	5·1	1·9	1·5	4·5
0·950	7·0	4·2	1·7	1·7	3·7
0·960	6·6	3·3	1·6	1·8	2·9

Inter-floor traffic

Floors above ground: 9

p_0	E_h	E_{s1}	E_{s2}	E_{ss}	E_p
0·650	9·0	17·6	0·2	0·0	38·8
0·700	9·0	17·4	0·3	0·0	32·1
0·750	9·0	16·7	0·6	0·1	25·9
0·800	9·0	15·6	1·0	0·1	20·1
0·820	9·0	15·1	1·1	0·2	17·9
0·840	9·0	14·2	1·3	0·3	15·7
0·860	8·9	13·2	1·6	0·4	13·6
0·880	8·9	11·9	1·9	0·6	11·5
0·900	8·8	10·5	2·1	0·8	9·5
0·910	8·8	9·5	2·1	1·1	8·5
0·920	8·7	8·5	2·2	1·2	7·5
0·930	8·6	7·5	2·3	1·4	6·5
0·940	8·4	6·5	2·2	1·6	5·6
0·950	8·2	5·1	2·0	1·8	4·6
0·960	7·9	4·0	1·9	2·0	3·7
0·970	7·3	2·9	1·5	2·1	2·7

Floors above ground: 10

p_0	E_h	E_{s1}	E_{s2}	E_{ss}	E_p
0·700	10·0	19·5	0·2	0·0	39·2
0·750	10·0	18·9	0·5	0·0	31·6
0·800	10·0	18·0	0·9	0·1	24·5
0·820	10·0	17·3	1·1	0·2	21·8
0·840	10·0	16·4	1·3	0·3	19·2
0·860	9·9	15·3	1·7	0·4	16·6
0·880	9·9	14·2	1·9	0·5	14·1
0·900	9·9	12·3	2·3	0·8	11·6
0·910	9·8	11·6	2·4	0·9	10·4
0·920	9·8	10·4	2·5	1·2	9·2
0·930	9·7	9·1	2·5	1·4	8·0
0·940	9·5	7·8	2·5	1·7	6·8
0·950	9·4	6·4	2·4	1·9	5·6
0·960	9·1	5·0	2·2	2·1	4·5
0·970	8·5	3·4	1·8	2·4	3·4

Inter-floor traffic

Floors above ground: 11

p_0	E_h	E_{s1}	E_{s2}	E_{ss}	E_p
0·750	11·0	21·1	0·4	0·0	38·0
0·800	11·0	20·1	0·8	0·1	29·5
0·820	11·0	19·6	1·0	0·1	26·2
0·840	11·0	18·8	1·3	0·2	23·0
0·860	11·0	17·8	1·5	0·3	19·9
0·880	10·9	16·4	1·9	0·5	16·9
0·900	10·9	14·6	2·3	0·7	13·9
0·910	10·9	13·5	2·4	0·9	12·4
0·920	10·8	12·3	2·6	1·1	11·0
0·930	10·7	10·9	2·8	1·4	9·6
0·940	10·6	9·4	2·7	1·7	8·2
0·950	10·5	7·7	2·8	2·0	6·8
0·960	10·2	6·0	2·5	2·3	5·4
0·970	9·7	4·4	2·1	2·6	4·0

Floors above ground: 12

p_0	E_h	E_{s1}	E_{s2}	E_{ss}	E_p
0·800	12·0	22·4	0·7	0·1	34·8
0·820	12·0	21·9	0·9	0·1	31·0
0·840	12·0	21·0	1·2	0·2	27·2
0·860	12·0	20·0	1·5	0·3	23·5
0·880	12·0	18·6	2·0	0·4	19·9
0·900	11·9	16·6	2·4	0·7	16·4
0·910	11·9	15·5	2·6	0·9	14·7
0·920	11·8	14·2	2·8	1·1	13·0
0·930	11·8	12·8	3·0	1·3	11·3
0·940	11·7	11·1	3·1	1·6	9·7
0·950	11·6	9·3	3·0	2·0	8·0
0·960	11·3	7·2	2·8	2·4	6·4
0·970	10·9	5·0	2·5	2·8	4·8
0·980	10·0	3·1	1·8	2·9	3·2

Inter-floor traffic

Floors above ground: 13

p_0	E_h	E_{s1}	E_{s2}	E_{ss}	E_p
0·820	13·0	24·1	0·8	0·1	36·1
0·840	13·0	23·3	1·1	0·1	31·7
0·860	13·0	22·4	1·5	0·2	27·4
0·880	13·0	21·0	1·9	0·4	23·3
0·900	12·9	19·0	2·4	0·6	19·2
0·910	12·9	17·8	2·7	0·8	17·2
0·920	12·9	16·4	2·9	1·0	15·2
0·930	12·8	14·7	3·1	1·3	13·2
0·940	12·7	12·9	3·3	1·6	11·3
0·950	12·6	10·9	3·3	2·0	9·3
0·960	12·4	8·5	3·2	2·5	7·4
0·970	12·1	6·0	2·7	3·0	5·5
0·980	11·2	3·6	2·0	3·2	3·7
0·985	10·3	2·4	1·5	3·1	2·8

Floors above ground: 14

p_0	E_h	E_{s1}	E_{s2}	E_{ss}	E_p
0·840	14·0	25·6	1·0	0·1	36·6
0·860	14·0	24·6	1·4	0·2	31·7
0·880	14·0	23·1	1·9	0·3	26·8
0·900	13·9	21·1	2·4	0·6	22·1
0·910	13·9	19·8	2·7	0·8	19·8
0·920	13·9	18·5	3·0	0·9	17·5
0·930	13·8	16·7	3·2	1·3	15·2
0·940	13·8	14·8	3·5	1·6	13·0
0·950	13·7	12·4	3·5	2·1	10·8
0·960	13·5	9·8	3·5	2·5	8·6
0·970	13·2	7·1	3·1	3·1	6·4
0·980	12·4	4·3	2·3	3·5	4·2
0·985	11·6	2·9	1·8	3·4	3·2

Inter-floor traffic

Floors above ground: 15

p_0	E_h	E_{s1}	E_{s2}	E_{ss}	E_p
0·860	15·0	26·8	1·3	0·2	36·2
0·880	15·0	25·5	1·8	0·3	30·7
0·900	15·0	23·5	2·3	0·5	25·3
0·910	14·9	22·2	2·7	0·7	22·6
0·920	14·9	20·7	3·0	0·9	20·0
0·930	14·9	18·9	3·4	1·2	17·4
0·940	14·8	16·7	3·6	1·6	14·9
0·950	14·7	14·1	3·8	2·1	12·3
0·960	14·6	11·3	3·8	2·6	9·8
0·970	14·3	8·1	3·4	3·3	7·3
0·980	13·6	4·9	2·6	3·7	4·8
0·985	12·8	3·3	2·0	3·7	3·6

Floors above ground: 16

p_0	E_h	E_{s1}	E_{s2}	E_{ss}	E_p
0·880	16·0	27·9	1·7	0·2	34·8
0·900	16·0	25·8	2·3	0·4	28·7
0·910	15·9	24·5	2·7	0·6	25·7
0·920	15·9	22·8	3·1	0·8	22·7
0·930	15·9	21·0	3·5	1·1	19·7
0·940	15·8	18·8	3·8	1·5	16·8
0·950	15·8	15·9	4·1	2·0	14·0
0·960	15·6	12·7	4·1	2·7	11·1
0·970	15·3	9·3	3·7	3·4	8·3
0·980	14·7	5·7	3·0	3·9	5·5
0·985	14·1	3·9	2·3	4·0	4·1

Inter-floor traffic

Floors above ground: 17

p_0	E_h	E_{s1}	E_{s2}	E_{ss}	E_p
0·880	17·0	30·1	1·6	0·2	39·1
0·900	17·0	28·0	2·3	0·4	32·2
0·910	17·0	26·7	2·7	0·6	28·9
0·920	16·9	25·1	3·1	0·8	25·5
0·930	16·9	23·1	3·5	1·1	22·2
0·940	16·9	20·7	4·0	1·4	18·9
0·950	16·8	17·7	4·3	2·0	15·7
0·960	16·7	14·4	4·4	2·7	12·5
0·970	16·4	10·6	4·1	3·4	9·3
0·980	15·9	6·5	3·3	4·1	6·2
0·985	15·2	4·3	2·5	4·3	4·6

Floors above ground: 18

p_0	E_h	E_{s1}	E_{s2}	E_{ss}	E_p
0·900	18·0	30·4	2·2	0·3	36·0
0·910	18·0	29·1	2·6	0·5	32·3
0·920	17·9	27·4	3·1	0·7	28·5
0·930	17·9	25·4	3·6	1·0	24·8
0·940	17·9	22·9	4·0	1·4	21·2
0·950	17·8	19·8	4·4	2·0	17·5
0·960	17·7	16·2	4·7	2·6	14·0
0·970	17·5	11·9	4·5	3·5	10·4
0·980	17·0	7·3	3·6	4·4	6·9
0·985	16·4	5·0	2·8	4·5	5·2

Index

Acceleration of lift cars, 67–9
Accidents, 14
Accuracy of calculations, 25, 34
Airport buildings, 10, 19
Auditoria, 19, 110

Bunching of lift cars, 25, 43–4

Car parks, 16, 19, 79
Circulation core, 12
Circulation cost, 3–6
Computers, use in design, 6, 21, 26, 47–51, 56
Control systems, *see* Lift components
Corridors
  disabled persons, 114–15
  fire escape, 107–9
  flow capacity, 92–100
  planning, 11–12
Courthouses, 19
Crowd density, effect on walking speeds, 95–6, 101
Cycle time of lifts, 22–3

Delays to lifts, 13, 25, 44
Department stores, *see* Shops
Disabled persons
  corridors, 114–15
  doorways, 117
  fire escape, 11, 106
  handrails, 116
  lifts, 84
  ramps, 116
  stairs, 115–16
Doorways
  disabled persons, 117
  fire escape, 106–12 *passim*
  flow capacity, 100
'Down-peak' traffic
  initial selection of lifts, 21
  lift performance calculations, 31–4

Education buildings, 19, 97, 110
Elevators, *see* Lifts
Erlangian distribution, 40–3
Escalators
  capacity, 59–62
  components, 87–9

Index

Escalators—(contd.)
  dimensions, 89–90
  planning, 9–10, 61
Estimate of building population, 109
Estimate of demand for lifts, 19

Fire
  access lifts, 84, 113
  access stairs, 113
  escape routes, 3, 91, 101, 105–17
Flats, see Residential buildings

Height of buildings, 1–3
Hospitals, 19
Hotels, 16, 19, 79, 110

Information
  to building owner, 14
  to users, 13
Interval between departures of lift cars, 16, 22, 25, 30–3, 41, 44, 51

Legislation
  disabled persons, 114
  fire escape, 107
Leisure buildings, 16, 19, 110
Levelling of lifts, 66–7, 70
Lift cars, required capacity, 25–6
Lift components
  buffers, 65
  cars, 72–7
  control systems, 43, 52, 78–81
  counterweights, 78
  safety devices, 13, 72, 75, 78
  shaft construction, 81–3
  suspension, 77–8
  traction machinery, 66–7
Lift lobbies, 8, 12, 16, 112
Lifts
  access for firefighting, 84, 113
  criteria for selection, 15–19
  dimensions, 63–5
  misuse, 13
  performance calculations, 22–35

planning, 6–8, 11–14
preliminary selection, 15–22
theoretical models of performance, 35–52
Lifts, electro-hydraulic, 65, 70
Lighting
  lift cars, 74
  machine rooms, 83
London Transport, 59–61, 100, 104

Maximum walking distance, 107–8
Mechanical failure, 13
Monte Carlo method, see Simulation
Motor rooms for lifts, 69–70, 83, 85

Negative binomial distribution, 41
Normal distribution, 40, 60, 92

Obstructions in corridors, 99
Occupant load factors, 109–11
Office buildings, 1, 16, 19–21, 107–13

Passenger conveyors
  capacity, 61–2
  components, 90
  planning, 10
Passenger movement time, 23, 133
Paternoster lifts
  analysis of capacity, 53–9
  components, 85–7
  planning, 9
Pedestrian lane width, 98
Poisson distribution, 25, 32, 38–57 *passim*

Queues for lifts, 12, 17, 42
  see also Waiting areas

Ramps
  disabled persons, 116
  energy expenditure, 103
  walking speeds, 94–5

Random interfloor traffic
  initial selection of lifts, 7, 21–2
  lift capacity, 27, 35, 39–40
  paternoster capacity, 9, 54–9
Re-opening of lift doors, 44–5
Residential buildings, 16, 19, 79, 110
Round trip time, 22–3

Sensitivity of calculations, 34–5
Shape of buildings, 1
Shops, 10, 16, 79, 93, 94, 97, 110
Simulation
  lifts, 17, 45–51
  paternosters, 56
Speed of operation
  escalators, 59–61
  passenger conveyors, 61–2
  passenger lifts, 66–72, 132–3
  paternosters, 86–7
Stairs
  disabled persons, 115–16
  energy expenditure, 103
  fire escape, 11, 109, 112–13

fire-fighting access, 113
flow capacity, 100–2
planning, 10–11
proportions, 102, 112

Transport buildings, 10, 19, 92–3

'Up-peak' traffic
  initial selection of lifts, 7–9, 18–21
  lift capacity, 22–31, 36–8
  paternoster capacity, 9, 53–4

Vandalism, 16, 76–7
Variation in lift cycle times, 40–3

Waiting areas, 12, 61, 103–5, 114–15
Waiting times for lifts, 16, 42
Walking speeds
  corridors, 92–7
  stairs, 102

Zones of lift service, 27–31, 43